Resistance and Redemption

Indigenous Environmentalism in the Age of Extraction

by Robert Lundahl

Table of Contents:

Chapter One

The Land You Know Has Been Given to You

I met Preston Arrow-weed along an empty stretch of highway in the Mojave Desert, at a place long ago inhabited by the Kamya and Quechan peoples.

It was along the shores of a formerly vast inland sea, to the north of a basin or sink that holds the now environmentally challenged Salton Sea, a shadow of its former self.

Preston sat on the front bumper of a Toyota pickup, looking out over a railroad track, at the former shoreline. I interviewed him for a film, then in the works, called "Who Are My People?"

His people had lived there. There were hundreds of small cairns, denoting cremations, the locations of the burials of people who once fished from its shores. At that time, it was the location of a proposed energy facility from NTR Toll Roads of Ireland (www.ntrplc.com).

The world had intruded on this indigenous landscape, as sacred as any. The knowledge gap was extreme, the listening gap insurmountable.

The Bureau of Land Management, which leases these lands to extractive and green energy industries, has a poorly understood mission. Once known as the General Land Office, its mission supported the goals of Western expansion and homesteading.

With the recent approval of Alaska's Willow, oil exploration project from ConocoPhillips, controversy over would-be copper mining at the native sacred site, Oak Flat, at Nevada's Thacker Pass, at the Salton Sea, and gold mines like Oro Cruz near Yuma, Arizona, that mission is on full display.

Particularly with the case of Lithium mining, as it is and would be, Berkeley Lawrence Livermore Labs suggests, though the potential impacts to the "green energy economy" may be hopeful, "significant technical challenges remain."

We begin this multi-part series on Lithium mining with the question, "Do we need to destroy the earth to save it?"

The Gold Mine

Robert Lundahl:
It’s really nice to meet you. Could you both introduce yourselves for whomever might be listening or reading, why are you here and where are you?

Preston Arrow-Weed:
Preston J Arrow–Weed, I'm on the Fort Yuma reservation. Oh, this is Faron.

Faron Owl:
Basically, I've worked with Preston and am just trying to help as much as I can on all these fights that we're involved with. It's been a very good experience.

Robert Lundahl:
I've been paying attention and that's what I wanted to talk to you about, so could you describe the event the other day? You guys had a run and went out to a sacred place with a geoglyph, and an official BLM geoglyph sign there.

This is close to Oro Cruz, which is a gold mine. Is that correct? What are the issues there with Oro Cruz? Why did you have that run?

Faron Owl:
Well, I took Preston over there. I went a couple times before Preston had gone, actually, to the Oro Cruz and American Girl Mines – which are close to each other. We went back there and looked at the areas of Oro Cruz, also, American Girl, and both of them are just devastating. Preston can pick up the story from there...

Preston Arrow-Weed:
That is not the only thing out there. There's more out there. It's been recorded by the BLM. They have many things out there, this is just one of them. There's more out there but they don't bring that up.

They know what they're doing, they should bring that up, but they don't bring up the research they've done. We've asked them to do it, to bring it in but they won't do it. So I believe they know what they're doing but they're still doing it. They still destroy.

Now I think too that they're important because eventually we might find the meaning of them (the geoglyphs) – because not even the experts know what it means. I don't think they do. If they do, they would say it, but I think there's so many things out there. We don't know even nature itself.

We don't understand sometimes what we're doing to it and I think that we see what happens when you do some things against nature and that's one of them. And one of the things they've done is destroying... like the mountain for instance and that I didn't like, the destruction of that mountain.

I don't like the destruction that I see simply to get gold to line their pockets to get money– that's not reasonable for me to say it's okay to go and just destroy everything. I don't think so.

I think when I and (other) people learned what their people saw and what had happened, but no one said nothing… when I saw it, well, I was angry because they were created the same way we were.

They created people. They created mountains. They're the same so that's when I was, I became angry, "let's do something about it."

I've even seen them talking about gold on television, telling people to save gold because that's a good thing to save and use later on, well you can save as much as you, well, as much as you can to save for your family or your future- but, what if the earth falls apart?

You know, I don't think you can use that gold to save yourself or to save the world but that's exactly what they're doing. They're digging for gold everywhere they can. For what? You know, what good is that?

Basically, for every sack full of gold that these people are getting, there's a place somewhere where they destroy the environment; they don't realize that. Somewhere creatures, animals are dying. They don't know that somewhere out there they could contaminate a lot of water. They don't know that.

As long as they have that gold in their hand, they think that's good, that's good for their future. There will be no future if you keep doing that with the gold [mine]. And among other things that happen with that gold mine is also the lithium problem.

The Lithium Problem

Preston Arrow-Weed:
Over here, we're dealing with lithium. It's another thing and I think that they don't care because right now there's the geothermal out there and plus the San Andreas Fault over there, and they're digging right near it.

If they can, they will go on until they can cause an earthquake. They can do a lot of damage and so they really don't care. They just want whatever they can get out of there.

Of course it's the public that lets him do it. The public is the one that's responsible.

Robert Lundahl:
Yeah, the BLM and the Department of Interior – so you're talking about they let them do it. You mean the federal government?

Preston Arrow-Weed:
The companies do it. You have to let them do it. I mean, of course you know that corporations have a lot of money. The people with the money are doing this and I think the people with the money are also running the government, you know...

They're doing that, so they listen to them and then the lowest one on the totem pole is the public.

The public wants it because they think they're going to get money. They do want money, they do get it matter of fact, and they give grants in Imperial Valley. A lot of people get those grants, so they've bought their way in there.

So those things happen but, I think even the public- they are not from this area. They came from somewhere else and if anything happens that it's no longer livable they can just pack up and leave. But us, we came from here.

We were here way before they came, so where can we go? We have no place to go.

All the creatures in the desert, they have no place to go either. They're going to die there, but it doesn't matter to the people, they want the gold, they want whatever they can get, the lithium, whatever.

There are so many things that they do without any consideration to the future, what can happen, what will happen. I think there are things that are happening in this world.

That they will not admit that there are things happening now, well one of the most famous things is global warming. You talk about that but the result of that are these storms and floods, and the weather is changing and everything is changing. That's because of what we did. I mean what humans have done.

There's no longer... there's no balance anymore. I mean if there's gold there that's meant to be there. If there's lithium there, that's meant to be there. Whatever is there is meant to be there.

Anything you, that you want to use, use a little bit if you need it, don't take all you can, you know, just like when we want to eat, we know that we can get meat but we don't try to get all the meat and then freeze it and sell it, no, we just get what we need and eat and go on. You know, in moderation.

When the Spanish came, they wanted to sell everything. We didn't think like that. We just get what we need, what we want and that's it. We feed our family, that's it. But when they came, oh, they wanted to sell it. They want to do all kinds of things and that's the way we are today, they are today.

They want to get as much as they can, all they can, but they don't think about the consequences of that.

Robert Lundahl:
I have a question on the lithium production. They want to tie that to the geothermal because the geothermal makes a brine and then they can extract the lithium from that. It sounds like a really good idea, like two processes at once, you know, to save energy. But you know, that's untested technology.

Preston Arrow-Weed:
I understand they really don't know what they're doing, too, I hear they don't know what's going to happen – It's almost experimental too, what they're doing, so they really don't know either, don't care I guess, if it goes wrong, they'll deal with it whatever way they're going to do it...

Somebody Pays

Robert Lundahl:
Well, somebody pays. Who pays? Who takes the risk? Is it you? Is it your community that takes the risk if they fail and you're left with a big mountain of toxic sludge?

Preston Arrow-Weed:
The community does not know the problems that we're having. They don't know. All they know is that you can get money and that's it, but the problems we have here, they don't know that.

I mean, what we're talking about like the environment, the destruction of the environment, things like that, they don't know that. They only want... money. All they want is, I don't know, comfort, I guess, so think about that.

In our time, our people, they knew that. For instance, they moved from place to place; when we depleted an area of certain things, we would move on. They would move on to another place once they depleted the area of our game and let it build up again.

Some people stayed. You can see in our history where we came from the North; we came down, we lived here and here and here all the way down. But when you came to this area for instance, it was underwater.

The shoreline was way up at Palm Springs and we were on the shoreline. We lived there and then when our water receded, we came down into Imperial Valley. Because they had archaeological studies done, they found artifacts also. When the water receded and they came down there were others who went along the mountains too. Some went to Jacumba, to the ocean, to San Juan Capistrano, up towards the mountain again by Palm Springs, then came back down south to Imperial Valley into the Yuma area.

By that time the Colorado River was flowing, but it took thousands of years for that to happen – science will not accept that, but then when I talk about it they say, oh yeah, yeah, we do have artifacts over there at uh Palm Springs... in that area coming down this way.

They say "yeah" but what does that tell you, that we've been here a long time. And it is our area.

Robert Lundahl:
You mentioned that you have fish traps that are similar to up north. Is that true? Can you tell me a little more about that?

Preston Arrow-Weed:
The Salton Sea, on the northwest side of the Imperial Valley now, fish traps were built there, along the banks. It's like a "v." They piled rocks and made a "v" right along the shoreline, that V is showing a little opening.

They would have a big circle and they would all start making noise and whatever and all the fish would come closer. Closer they got into that hole and went in there and then they closed it.

That's how they got their fish. And there's a lot of them over there on the West Side, but of course it's long ago so you can barely see them but they're there because I've seen them too. They're there, it goes all the way up on the North side, and over there you could even see where the water level used to be along that mountain long ago.

Robert Lundahl:
We were talking about moving as opposed to migrating, right? And we talked briefly about the DNA evidence about migration from Alaska and they've dated that in certain cases, so are you talking about migrating or moving?

Preston Arrow-Weed:
I can only tell you that my people in this area do not migrate, they just moved. They stayed until they depleted the area, or they needed something else, so they moved to another place, another, there's evidence of that.

They might stay two, three, four, maybe thousands of years and then they moved, but it took all that time to reach where we're at– so you can see how long we've been here based on that.

You can see that we didn't migrate here, if we did migrate, we would come straight down through here bam, bam; it doesn't happen that way, but we did from place to place until we got here.

Robert Lundahl:
The Lightning Song that you sing tells that story, right, because you sang it out there on the land – it's part of it. You got everybody singing the first couple of lines I think...

Preston Arrow-Weed:
Oh yeah, I've used the Lightning song a lot. Our journey down this way when we first lived here, but there again it's the same thing I'm saying... to migrate, you can migrate in one day.

The geese fly through here, could fly here through the one day in a migration but I don't think we did that in one day. It took thousands of years, hundreds and thousands of years. That word they use, I don't think it fits us at all.

I heard that some people went to court, Natives, Wendsler (Wendsler Nosie Sr.), Oak Flat, and the people who wanted the land told him that that land belonged to the government.

All right, how did they get it? When you really look at it, they stole it. You know I talked about our land, and they say that BLM has it now. How did BLM get it?

I noticed like for instance the gold mining, they say they're using claims, claims that they bought from someone who was out there. Their claims are only 200 years old. The oldest claim… there's evidence we were there before that claim. But that doesn't mean nothing.

The United States declared that, you know, that's "not enough proof" so they go by their claims that they create or what they put there.

Robert Lundahl:
If that's not right within their memory, within their organizational memory of the United States of America…

Preston Arrow-Weed:
You know they could make any law to say that this is the way it's supposed to be and then whatever you know is gone.

They're doing the same thing using religion. Our religion was before the Bible, I say. Our religion goes way back before they did that.

I asked about the religion once, the Bible, and they said something like two three thousand years or something like that. Well we go beyond that and what we believe, so who is telling the truth?

I mean they can make any law to say that, like for instance, in the state of California, they wanted the land, they wanted the gold, and what did they do?

They started shooting and killing the Native Americans in California and then to make it worse, they enacted and passed laws that there was a bounty on them, and you can shoot them and kill them as long as you bring them their heads. See, they made that law– see, they did that too.

They've done everything to get what they want on this land, and is that justice?

Is That Justice?

Preston Arrow-Weed:
I mean it's wrong. I even kind of think about today. Recently I heard that Russia is committing crimes against humanity. America has been doing that since the time began for us since they came, they've been doing the same thing.

He came and took it; still, they provoked our people into war, but we couldn't win the war because of the modern technology they had at the time up to today. We couldn't win and they took our land – they even enacted laws, enacted rules, enacted everything and we didn't understand it either. Do you think that's fair?

I think the Russians probably learned from the American people how to do it, so it's been done. America isn't as great, you know, in what they've done.

Preston Arrow-weed and Faron Owl (Quechan)

Robert Lundahl:
There are all these projects that have something in common. We are talking about the Willow oil exploration project up on the North Slope.

It's a traditional area and people live subsistence style there and that's where they gain their foodstuffs, so basically that means, what, you have to go to Safeway now? And then you went up and talked to the Nevada mining people, the indigenous mining people, about Thacker Pass, right?

And that's another lithium production area, where they hope to produce lithium, but it was on the site of a massacre that the BLM claimed not to know about as the project was approved by the Trump administration 5 days before leaving office.

"Thacker Pass is spiritually and culturally significant to tribes in Nevada, an intrinsic part of their history and stories, says the tribe (Reno Sparks Indian Colony). It's also one of the few remaining places in the Great Basin where tribal citizens can still gather traditional foods such as chokecherries, wild potatoes and onions, and medicines like the toza root." -Nevada Current

And then you have Oro Cruz about gold mining, and I don't know what chemicals they use there but it sounds pretty toxic to me (cyanide, cadmium, arsenic). And then the question of Lithium Valley, where is Lithium Valley, you know? I mean where does it start? Where's the boundary? Whose land is in it?

No one knows that and they've never done it before (extracting Lithium from geothermal brine), which Lawrence Livermore Berkeley Labs cautions involves "significant technical challenges."

"Geothermal brines in the Salton Sea region of California are expected to be a major domestic source of lithium in the future but that significant technical challenges need to be overcome."

Preston Arrow-Weed:
I don't know either because I've heard of the Lithium Valley thing. I think they have a board or something and our representative isn't our greatest and I mean he's not saying much. I saw them once at the meeting, but nothing was said about it of course. I complained a lot.

Our belief is just as strong. There's a legendary story about the area where the ancient Salton Sea was.

Our belief is that a giant snake came in there, to Imperial Valley, from the ocean. It was a white, big giant snake with black spots. The snake was a repository of knowledge. It came into Imperial Valley and surfaced in the ocean there somewhere. I would think it would probably be Obsidian Butte.

They had a ceremony at what they call the House of Darkness; they built a little house there and the snake went into it. They closed him up and he couldn't get out and they burned it, like cremating him. And when it burned, all the bodies popped from the fire and all the black spots, the repositories of knowledge, left the body, went out and went to all the mountains and there's some mountains around that area that it went to... there's one south of them called Eagle Mountain, and all to the north and all the way to the west, and Oro Cruz, some of that [knowledge] went into those mountains too.

That's what we believe, what the Kamya believe in, the Quechan, but that doesn't mean nothing to them, you know, they expect us to believe in the Bible (laughs), so things like that happen and I think it's wrong to even consider that, that their Bible is better than ours.

Our belief doesn't mean nothing, but it shows that our creator does not want to destroy everything; according to their belief it's okay to destroy, so which is better, you know, which is better? But then on the other hand I do believe that there are some good Christians too. I believe that's true- but they're not around. People are using religion to do the wrong things, too. I think it's probably a good book. I've seen some of it. I don't think it's bad, but people are misusing it to do what they want to do.

I did research when I was in college on Jesus Christ. There was a man, he did exist, he did go talking about the things that were wrong, and he tried to stop it. He talked about the things that they were doing and because of what he said, what he was doing, they killed him. They killed him because of that.

There was such a man as Jesus Christ; there was a man, it's true, and they killed him because he was telling the truth, and he was trying to stop it. I think he was trying to say the Bible was okay, if you did it this way it's okay but that way's wrong... but they wouldn't listen to him, they tried to change it.

I'm saying that all religions are okay but it just depends on how you use it and I believe- we believe that there's different ways to reach the Creator.

Our way is okay, their way is okay, everybody's way they have that's their way to reach the Creator, no one is better than the other. But America thinks theirs is better than ours, see that's what happened.

Robert Lundahl:
Okay. Well, that's religion. Let's talk about law. There's a bunch of laws here that nobody seems to pay much attention to these days with all these projects, like NEPA, National Environmental Protection Act, the National Historic Places Act that protects cultural resources.

NEPA has a provision for the human landscape– everything people do in the cultural landscape, however far that extends – so we have these laws, what's the problem?

What's the problem with... what happens when these projects get into court, and they don't listen?

Preston Arrow-Weed:
They change them but it's not fair the way they do things, not fair the way they fight or whatever, it's not fair they got the upper hand, so that's what's happening to us.

Who will fight for us, who will fight for us? We, at home here, at this point there's only about a handful, they're saying things, doing things what they can. Faron and I, and some people who've joined us but the rest don't know.

Avi Kwa Ame

Preston Arrow-Weed:
I try, we try to tell them, what to tell everybody, what we know. We had a gathering here last week, Saturday. I thought it was very successful. And we just returned from one in Parker, and I thought that was great. There was a lot of people there.

Avi Kwa Ame

Then recently Avi Kwa Ame, the National Monument. See, we got that and so Avi Kwa Ame is what they declare the National Monument. Our home here or our paths and roads go directly to that mountain, so we're connected to it.

So, what is America going to do? Cut us off? Disconnect us from our mountain? Because that's our belief too, that's our belief too, we sing the same song, and we tell almost the same stories about that area, so America to cut that belief, you know cut whatever we talk about, oh that's really something to do that.

So we are connected to that now, and that's going to be a danger to them. They can try their way to stop that too. Eventually. They can probably throw it out, say "hey yeah, forget that" you know. They'll do that too. Whatever they do, they'll change. It doesn't matter.

All they have to do is have a quorum and they vote on it and they say okay, forget it. To me that's not fair. They're not going along with what is fair and what is right or wrong; they're not going by them.

Robert Lundahl:
Well you know some people will say everything's changing now. There's no thought of co–management it's all the old way like "we want to do what we want to do" and you guys don't really get to have much input into this process and if you do, "consulting" whatever that means you know, get somebody on the phone or try to call or something it's "consulting" but there's not "consent."

Preston Arrow-Weed:
No, I think that she's under pressure from all sides, you know, every side. You're under pressure to do this or do that; if you do this, this is going to happen.

They got it to where she can't say yes or no. Because, in little politics just little politics- I was a tribal councilman one time... two terms I served and I know the pressure. What's going to happen if I change? or, Do this!

You know, they had me, but I just made a bold move "nope it's going to be that way" and that was it- what I said went. I went the right way and I don't care about the consequence, we're going to do it! And that's what I did, but I can see why leadership is kind of hard.

They don't want to do this because they're this or that for their own safety or whatever. I see that and I'm sure that happens to the Department of Interior.

And we're talking about money – if you do this, we're going to cut your people off or whatever.

I once had a federal officer come to investigate a case somewhere. The county and everybody scared her, and she took off – she dropped the whole thing.

Robert Lundahl
Thanks for commenting on that because you know when you read the papers it goes one way and then it goes the other way so it's good to have, it's good for people to have an understanding that even at a top level that native people are getting pressure.

Preston Arrow-Weed:
Well you know I'll tell you what happened years ago. There was another Secretary of Interior, I forgot her name, and she was very strong in a lot of things. The Native American didn't like her.

Well, they had a position on the Desert Advisory Council in Imperial County and that area, so at that time I was living in El Centro with my wife, and my name came up," Preston J. Arrow-Weed, El Centro".

Grapevine Canyon

So, everybody wanted me in there, and everybody voted for me. The environmentalists all wrote their letters to get me in there and I think she looked at it and Preston J. Arrow-Weed living in El Centro now she probably thought I was a wanna-be Native American.

So, she put me on, that was her mistake, but I got into the Desert Advisory Council. Boy, when I went to those meetings I just went against everybody that came in.

I was the only one and nobody would side with me but they all respected me. They said, you know, Preston is the conscience of the board. Then I said, "Why the hell don't you follow me then?" I said, "You know this."

I don't care how big they were. I don't care if they were very well known, whatever. I said no. I said, "you're wrong." I would argue against them.

I did that for three years. I wouldn't give in, but nobody gets that chance too often but I had that chance and I did it; what I could do. I never accomplished anything because the odds were against me, the same thing right now, right now.

You know, Faron and I are doing this thing and if we lose and they beat us, it just doesn't mean that we lost. It's... they lost. They lose- the people lose, everybody loses, every creature in the desert loses, everything that's out there loses because nobody listens to us.

We don't lose. I think Faron and I we both have fine ways to go somewhere else and be safe if we wanted to be but it's the people here- they don't know, even the tribal council doesn't know, that if we lose, they lose too, and yet they're the leaders, you know, and the county, the people living here, they all have to go somewhere but things like that happen, if we lose they lose.

Right now, they're fighting for the money, that's what they're doing, and it is our land.

We were here before. There's history, too, about Imperial Valley that when the [white] people first came to the valley, they did try to drive out all the Kamyas that lived here. They drove them out, shot them. They killed them. They started wars against them, they did that too, and eventually they did get them out of there.

So, they never did anything right, they were all wrong, what they've done. So now they're going to destroy you too. Now they got rid of the people, now they're going to destroy the land, they can destroy everything in their religion...

Robert Lundahl:
Well Imperial Valley, because of the Salton Sea and its problems, there's a lot of health problems in Imperial Valley, right? And very poor air quality. They can't get air quality, Clean Air Act funds for the county and that's a pretty bad situation, so if this goes bad, you know, then you have a layer on a layer...

Salton City

Preston Arrow-Weed:
Let's go back to that. Now, a long time ago, they say that the Salton Sea went down and dried up and it came back, again it dried and came back but then when it dried up, nothing

happened. But since the people came to Imperial Valley, all the chemicals, whatever, everything went in there.

They're the ones that damaged it, it wasn't us.

If it was the natural way, it would have been okay but even the natural way was destroyed by farmers using these chemicals and fertilizer and other things. The cyanide went in there too, underwater.

Yeah, so that's them that did it. So they're saying if the Salton Sea goes down, all that [poison] is going to come up, and the wind blowing the air this way it's going to affect us.

So, they want to do something down here, they're going to get all that dirt and cover it up... see, that's what they did. They did that.

The natural way would have been better if it just dried up and if they had nothing in there, it would have dried up and then maybe built itself back up again.

Young Elders

Preston Arrow-Weed:
But hopefully we'll keep trying, and the young people are coming to our classes, you know we call them young elders. There's no more of the old Elders anymore. We call them young elders because we're trying to teach them all we can.

There's an old legend that came back to me:

I don't know if it's a warning or what, but I heard that an enemy attacked the people. The enemy came in and killed every one of the tribe. The young kids all ran and hid.

After they were gone the young people came out and they were trying to survive and this old Warrior that had been gone for a long time, showed up and he asked, “What happened?”

They told him. Then that old Warrior taught them how to survive and they survived, and he also taught them how to fight, and when they grew up and they all got bigger and bigger they went back looking for the enemy and they destroyed them, and they came back again.

That's a story, so here I am.

Robert Lundahl (To Faron Owl):
What are your thoughts as an educator? I mean Preston's getting his side of it and, you know, you have students that come into your class every day and you have to look at them in the eye and how does that go for you? How does it go for them?

Faron Owl:
Everybody that we've associated with from the Salton Sea to Fort Mojave, all of them are in that same concept that they need to get the young people involved, because it obviously, it's keeping the land the way it is, it's saving as much as they can and the sacred parts of it that we use, and that we've used, the plants and the animals that are out there.

Faron Owl gestures toward the mine

They have that approach of educating the children, getting them involved, culturally, and telling them how important this is, especially like Avi Kwa Ame, on how important it is to not just one tribe – it's seven tribes that are involved with that.

It tells that story from that beginning where if you go to the lithium portion of it there in Southern California, ours is gold. You go to Arizona; the Apache are fighting with Oak flat. You go to Nevada, they're fighting lithium; you go to Utah, they're fighting uranium.

Leadership Group

Faron Owl:
And so, all those ideas that people are coming into, for instance, hey we want to bury this or we want to dig this here or we want to see how we're going to create a green environment using two benefits (Lithium and geothermal).

All of that needs to have our people get together and create some type of, if you want to call it a network, or if you want to call it some type of joint effort, all of us sit together and come up with a plan.

Come up with goals and come up with a leadership group that can actually take charge and inform their people, let them know because, in our case, there's a lot of people who don't even know what's going on.

There's a lot of people that haven't been twenty-two miles from here just to see the impact of the gold mine and what it really does. So, I think once people see that, they realize that this is bad.

When we ran out last week to the geoglyphs, you know that was a story, and Preston met us out there and he told everybody the story.

You saw the film on that, and it shows how important our area is and Oro Cruz just a little ways and American Girl it's just right over that mountain, and how devastated the land is and a lot of people have not seen it because it's kind of hidden from the main road.

In the education portion of it, I think we're going the right direction getting our young people involved, culturally but more importantly about our environment.

Now every little thing that happens to it, how it impacts our environment, what we breathe, and what's most important for my tribal lands is that we are trying to preserve everything for our young people.

Robert Lundahl:
...mm-hmm and you said preserve “everything.”

Faron Owl:
Yeah, as much as we can… the language, the songs, the stories, and like Preston, he knows the songs, and he knows the stories. He knows the words to tell the people what's going on.

At this point we're in the beginning stages of creating something that's a positive for our community.

Chapter Two

Oak Flat: The Story Behind the Story
Mahlee Yellowfeather, Great Granddaughter of Geronimo

Mahlee Yellowfeather was raised in a back canyon, speaking only Apache. She was taught by the Elders.

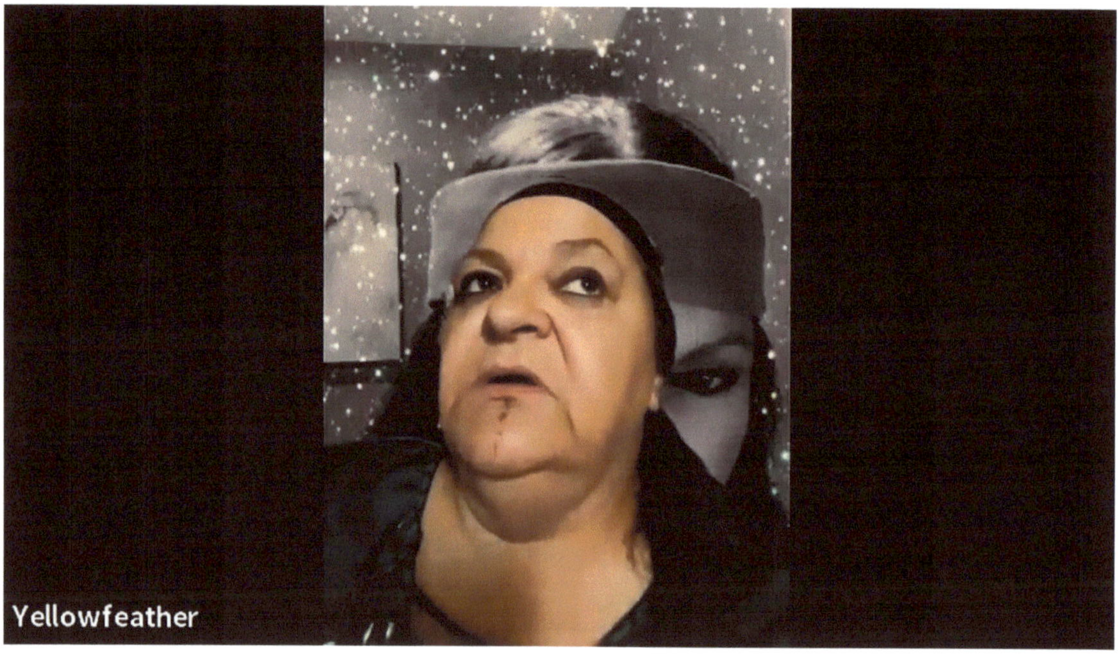

Mahlee is the great granddaughter of Geronimo. Geronimo's raids and related combat actions were a part of the prolonged period of the Apache–United States conflict.Here she tells of the occurrences at Oak Flat, and the meaning of the area to the Apache people.

Sometimes stories come to you. My curiosity had been piqued by the contradictions implicit in what we call "Green Energy," mining cobalt and lithium for electric vehicle batteries, to save the planet by digging huge pits, the depth of several empire state buildings; animals, people, laws, history, water quality, and ecosystems be damned.

It dawns on this writer we are witnessing an era of raw power plays, unmitigated greed, inequitable development, and in at least one case, sheer chaos.

At least we're going to test that assumption and hope better permitting processes prevail. The "sheer chaos" is best exemplified by the alleged history of Oak Flat, Arizona, where reported raids by international mining enterprises meet Chiricahua, Apache resistance over the development of what has been called the third largest copper ore deposit in the world.

I did not anticipate having a privileged glimpse of the life of Ghailee Geronimo.

Mahlee Yellowfeather is the great granddaughter of the legendary warrior. To her, Oak Flat is a sacred site. She tells the story of how one day, that sacredness was interrupted by an unexpected military incursion.

Illustrating the monetary value of her traditional lands to outsiders, there is an as yet unresolved dispute over law, land, and the right to mine in what has become no less incendiary a conflict of values than an old-time Western land war, continuing the Apache-United States conflicts of the 1800's, from before the West was "settled."

According to Mahlee Yellowfeather, the Chiricahua Apache want peace and are moving forward as a people, into a perhaps more powerful and effective position in the modern world of 2023.

The Flat Oak War

Robert Lundahl: What happened exactly, what did you see? When did you first realize that there was a problem that your family had to be involved in?

Mahlee Yellowfeather: We had the Flat Oak War. This is where we lost a lot of our members of our family. That's when they came over and they were trying to, well they started bombing us.

That was the really worst one. There that was the first time that they tried to take and take over the mine, and we lost a lot of people there, a lot of family, a lot of relatives, a lot of people in the tribe...

Robert Lundahl: And when was that?

Mahlee Yellowfeather: This was, oh God, it's been like- 20 years ago.

And they say that we inspired Standing Rock, but I don't see where we inspired Standing Rock because we didn't hold signs up in prayer, we fought.

The government was called, and they didn't deploy till after uh 24 hours while boots were on the ground.

By then we lost biggest part of the tribe.

They had helicopters, tanks, everything, so how do you not know, how does the government not know that they were coming across into the waters into the land to attack us? They did.

Robert Lundahl: I see.

Mahlee Yellowfeather: And the government waited to come and assist us. By that time, we already took them down, but we lost so much.

It was a bloodbath.

The Bloodbath

Robert Lundahl: Who were you fighting?

Mahlee Yellowfeather: It was people from Australia.

Robert Lundahl: Oh really. So you're talking about the miners. You were fighting the miners.

That's what's going on in Brazil with the Yanomami people and the mining, the miners come in and they come into the indigenous land, and they do what they do.

So that was Rio Tinto?

Mahlee Yellowfeather: Yes.

I just don't publicize it. We don't publicize it. We don't really even talk about it too much due to all the relatives that are not here any longer.

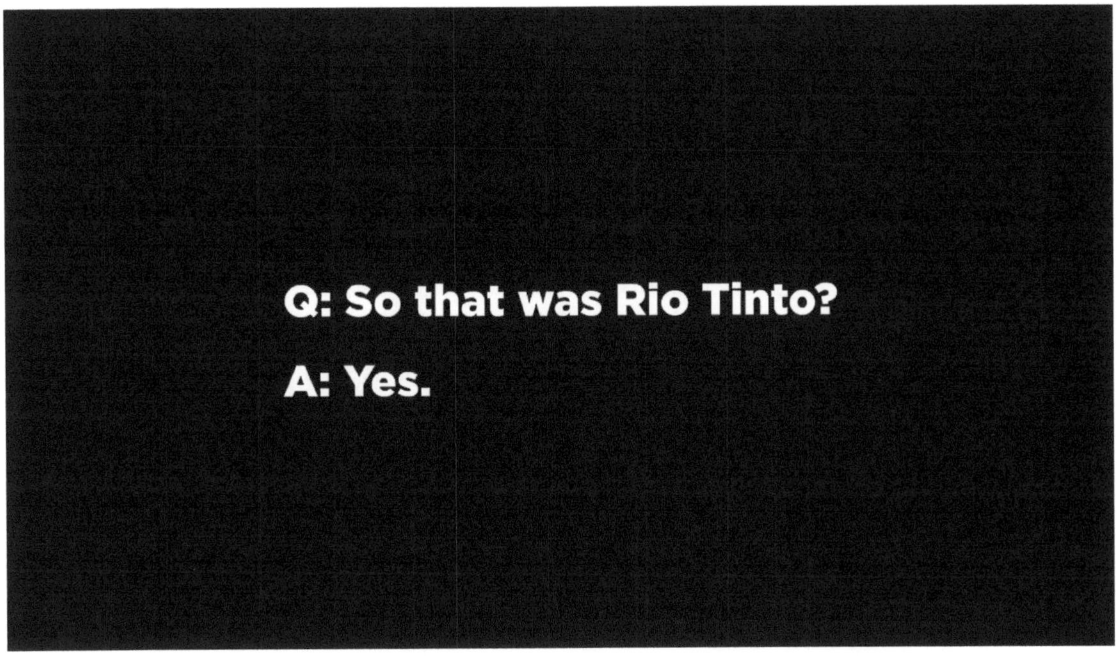

Us Apache, you know, are very secluded. It was not a very good day. We were all sitting and in celebration ceremonies and next thing you know, we're getting bombed.

Robert Lundahl: What gave them the right? Did they have any right at all?

Mahlee Yellowfeather:

From what I understand, the dude had a deal from the 1800s. It was a mining pass from my great-grandfather, Ghailee Geronimo, to his great-grandfather stating that they can mine for a year, and he thought, well, he could take it and run with it again... and it's like, no...

Robert Lundahl: Right.

Mahlee Yellowfeather: No.

Because he would let some people come over and mine, but it wasn't like a whole bunch of miners, you know, it would be like, you know, “okay, oh yeah, you can go ahead and mine for a little bit,” if it was good for the people.

But these companies coming over, they just want to suck them dry. And this is what this guy, (what) his intentions were, and he thought, “Oh he found a piece of paper that his great-grandfather had, that he thinks he could pull (that) and say, “I'm entitled.”

No, you're not entitled to something that was between my great grandfather and your great-grandfather, and that deal is over.

That Deal is Over

Robert Lundahl: Were you involved in discussions with the Department of Interior at that point, or the BIA, or the BLM, or anyone like that, or were you just on your own?

Mahlee Yellowfeather: It was our elders that were involved with the discussions at that time. I was just involved in the fight, in protecting and fighting with my family.

Ghailee Geronimo

Robert Lundahl: I see. So where do things stand now? Flash forward to the future a little bit from that point.

Mahlee Yellowfeather: Well, we're not going to let them have it, so more likely I would say that we're going to go back to war.

It is sacred grounds, and you're not entitled to it. I mean I feel that these people already took enough from everybody.

Robert Lundahl: So does the BLM claim that the government owns it?

Mahlee Yellowfeather: The government thinks they own everything, in Department of Forestry, and everything else, and really they don't.

The Contract

Mahlee Yellowfeather: There are contracts that were put out back when Teddy Roosevelt was around, and that's why we stand our ground.

He gave those titles to, uh, Geronimo, which we all possess; there's no government with "D" course of title. We have "D" course of title, so how are they taking possession of something that don't belong to them?"

Robert Lundahl: Short memory?

Mahlee Yellowfeather: And that's like they think they know the story of Ghailee Geronimo, which is not true either. He had a twin brother, it was Gugalah (phonetic), which "thy brother betrayed me."

Robert Lundahl: So they don't have a right... this is the federal government I'm talking about... they don't have a right. They don't have ownership. You never gave them ownership.

You gave a company the right to mine for one year, from Geronimo, your relative.

And you're afraid, you're saying, that maybe you're going to go to war again, or it's going to be a battle in some manner, right, so what does it look like today? I mean, this is in the news today, and people are trying to become aware of it–you know, we're trying to make people aware of it, so what do they need to know?

We Want Peace

Mahlee Yellowfeather: We want peace. I mean we want peace; we want to be left alone. We want to be able to live in peace and be left alone.

We don't want our stuff messed with, especially our sacred grounds. Our people are there. That was the grounds of massacre, of where families were massacred, and that's what makes it more even sacred.

Oak Flat

Most of those mountains up there, through there, are just sacred grounds. It's just kind of like Mount Rushmore up in there. Those were all sacred grounds where we used to take off from Four Corners... (with) our (deceased) Warriors and Chiefs and bury them up there, you know, and then they just take it and (say) “Hey it's ours now.” That's why we're still in court with them over that.

You know it's like we have to have some type of, “Hey we're human,” you know.

Robert Lundahl: Yeah, it's amazing. 2023 and we're still here.

Mahlee Yellowfeather: We're still here. We're here, and we're still repeating the past.

Robert Lundahl: I was just curious, so if you have burials on the land there are various laws...

Mahlee Yellowfeather: Yes, there are.

Robert Lundahl: So, whatever deal they think they made, or whatever they think they have, there are still laws that say you can't do this because of the burials and because of the cultural resources (present there).

Mahlee Yellowfeather: Exactly, we don't use the mine to go make money. It's not about the damn mine, plus we don't want to suck Mother Earth. We're trying to heal Mother Earth.

Robert Lundahl: With all this mining, lithium, copper, gold, everything…

Mahlee Yellowfeather: Gold. Yeah, the gold is the number one greed; (that’s) why you're not going to see natives wear it. It's a blood money, right?

Hoka Hey

Robert Lundahl: Can you tell me a little bit about your family? You mentioned that you're related to Geronimo, and some people know that story and some people don't, you know, in the public, but he never gave up, did he?

Mahlee Yellowfeather: No, he did not, and he would always say "Hoka Hey," it's always a good day to die.

He did not die in Fort Sill. (He) actually died right there in Skull Canyon at the base of the tree, the same tree that he pulled his horse up and he took his long walk.

Do you think they're actually going let the fiercest Native run the lands, go here, there, there, carrying guns, steal?

I mean you could go to my Pinterest and see all his pictures from the age of youth to the age of hitting 99 years of age.

And the difference between him and GugalahGugalah (phonetic), he don't have the scar.

Geronimo had this scar (on his right cheek) he got that at the age of 11 by a bullet graze.

Oh yeah (Gugalah) is the one that bond (ed) with the Buffalo society, which is the Lakota, which was Red Cloud's son.

They are the ones that betrayed, with the Calvary and were slaughtering all the Natives within their tribes; they're the ones that slaughtered Sitting Bull and whole tribes, so when Red Cloud heard of it he put his own son to death.

Robert Lundahl: I see, so you're saying that he capitulated.

Mahlee Yellowfeather: What led him (Red Cloud) to that is being honorable to his people of the betrayal of his son.

There is a journal that was written in Ghailee Geronimo's own hand that we possess as well. My stuff was not Googled, my stuff was, I was born with it in my lap. Just like on my other side of my bloodline is the Cherokee.

I know that the land that Teddy Roosevelt gave Ghailee Geronimo (was) to stop the massacres of everybody and he cut the turtle's tail, that's why the turtle's tail is not complete for Turtle Island.

Theodore Roosevelt

The only way the turtle's tail will ever be united as one is when the last of the Totem, which the Apache, the last of the Totem Nations, unite into the four corners, we will all come together, and this is when we retrieve unity within all tribes.

That's including Canada. Canada used to be free to the open. We used to migrate all the way down, all the way up. All the way down, all the way up.

Robert Lundahl: You used the word migration, so was that over long periods of time, or was that ceremonial?

Mahlee Yellowfeather: It went with the food. Ceremonial, hunts, everything. It was different with the powwows back in their time; it was total unity with the different tribes.

We would have our powwows with different tribes, not outsiders, just within tribes and it would go from different tribes (to) different tribes, and there would be many tribes coming together.

Come in a good way, go in a good way. We would do different hunts, but we would migrate within the food, where the food goes, we went.

Robert Lundahl: I see. What food sources are you talking about? Are you talking about buffalo? Fish?

Mahlee Yellowfeather: Buffalo, fish, yes, everything. They followed the salmon just as much as they followed the buffalo.

Robert Lundahl: Wow, that's covering a lot of miles.

Mahlee Yellowfeather: They followed the horses, the wild horses as well.

Robert Lundahl: So that history is something that, you know, for people that watch a lot of television and Internet, and the modern youth, you know they probably don't quite get it. How do you teach your youth?

Mahlee Yellowfeather: We usually tell them in our story times during powwows or Sundances, gatherings, we tell them. The elders do teachings. I was raised by the elders, you know far back in the canyon. I was learned by the really, really wise elders.

A lot of the children nowadays they have lost their way to even their culture, which is sad– and then not to mention the people and their addictions.

They lie upon the genocide because we're the Seventh Generation; there's a lot of pain and suffering within the tribes, within the people, and this is all over.

But my elders on how, as Apaches, we feel– this is 2023. And a lot of us feel that we cannot sit here and evolve within the pain that we did not endure within the points of the spear. That was not for us to endure. It was left for us to remember and use as strength. Other tribes they swallow upon the pain and use it as a weakness and that's why they're stuck. We move forward.

We Move Forward

Mahlee Yellowfeather: That's why a lot of our people have already went to the people's schooling in Engineering, working for Lockheed, building the NASA, working for major medicines and hospitals, and evolving with their people.

You know- because we can't we cannot live here, not without evolving with them.

Robert Lundahl: What role does Oak Flat have in keeping things together for your tribe and for your people, for your family?

Mahlee Yellowfeather: For our family and ceremony, we stay in prayers. We stay in our blessings. We keep our ceremonies strong.

We don't really discuss ceremonies, but I will put this much, that we do have our Warrior ceremonies; we do have our Sundance ceremony–which our Sundance type is for when the girls become their women.

We also have our Ghost Dance. We have our spiritual ceremonies. Just about every day the medicine moms they go, they smudge within the land.

Robert Lundahl: And Oak Flat, so do some of these practices happen there?

Mahlee Yellowfeather: Absolutely.

Robert Lundahl: Regularly? So, it's ongoing and this is like life, and the way it's always been, you're not giving it up, and somebody's just trying to take the land out from underneath the culture?

Mahlee Yellowfeather: Absolutely, it is a stronghold part of our culture that we use all the time as part of ceremony, as a remembrance, and keeping. No. We can't let it go.

Chapter Three

The Fight to Save Thacker Pass

Power concedes nothing without a demand.

The times offer convulsions of opportunism, greed and power. Perhaps these signs are the remains of a Trumpian legacy, in mindset, as well as resource exploitation. In a world of "if you want it, take it," there is plenty of help. It seems there is no more graphic illustration available than the fast track permitting of mining operations for lithium, gold, uranium, and copper.

What money and mechanization bring is the destruction of entire mountains, extraction pits thousands of feet deep, attacks on community, and the integrity of culture–thousands of years of place-based peoples successfully living in a more harmonious balance.

Humboldt County Sheriffs arrive at Ox Sam Camp

Robert Lundahl: I'm going to ask you to introduce yourself just so we have it recorded, and we have your proper name and how you like to be identified...

Will Falk: My name is Will Falk. I am a co-founder of the organization Protect Thacker Pass.

We went and set up a protest camp on January 15, 2021, the same day that the Bureau of Land Management issued the record of decision for the Thacker Pass lithium mine.

We set up a protest camp where the open pit for that lithium mine would be in Thacker Pass, which is a beautiful mountain pass in Northern Nevada.

I am also an attorney, and I represent the Reno Sparks Indian Colony and the Summit Lake Paiute Tribe in litigation against the Bureau of Land Management for permitting the Thacker Pass lithium mine.

We've now filed two lawsuits, it all revolves around the Bureau of Land Management's failure to consult with any Indian tribes about the Thacker Pass lithium mine.

Robert Lundahl: Just as we're discussing. That seems to be very common. My question about "what do all these large projects, the large mining projects have in common"–I think that's one of the things they have in common, am I correct?

No Such Thing as Consent

Will Falk: Yeah, and I think it's important to understand the limits of the consultation under American law.

I think a lot of times, Americans thinking that they live in a democracy feel like if there's going to be a massive project cited next to them, they should have some say over whether the project happens or not.

Well, when we're talking about mining on public lands in the United States there's a law, the 1872 General mining law, which makes it so that if a mining corporation locates valuable minerals on public lands, then the federal government has to give them permits to mine. This is not a discretionary thing.

If there are valuable minerals there, the Congress has said that the highest and best use of American public land is mining.

Ox Sam Camp with Tipi

That trumps Native American spiritual uses. There have been numerous examples of literally the most sacred places in the world to various Native American communities being destroyed because of the 1872 mining law.

Robert Lundahl: Does that law give them title, or does that give them the right to mine?

Will Falk: It doesn't give them title. The government retains title to that land, but it does give them title to a mining permit. And when I say the "right to mine," if you would go out there and interfere with their mining operation, they would be able to call upon the full power of the State to remove you from that area.

Like I was saying, I think a lot of Americans think that they should have some sort of say, like the Thacker Pass lithium mine would be sited between two primarily agricultural communities, in Orovada, Nevada and Kings Valley, Nevada.

It's about, I think, 25 or 30 miles as the crow flies from the Fort McDermitt reservation. So you know in our "politics 101" mindset, we think that if all of those communities were opposed to this mine that they should be able to get the mine project canceled.

That's simply not how the law works. There's no such thing as consent under American law for projects like these.

Paiute Elder Josephine Dick with Dorece Sam

What the law does require government agencies to do is consult with the public and federally recognized Native American tribes about these projects.

We should understand that the definition of consultation is not a commonsense definition. An example I like to use is you have this old dying tree that is growing on the property line between you and a neighbor.

It's sort of split right down the middle so you want to take the tree down and being a neighbor, you would go and knock on your neighbor's door and consult with them, ask them if they want to take that tree down, and you would actually take your neighbor's concerns into account.

And maybe if your neighbor loves that tree and doesn't want to take it down, through the consultation you decide you're not going to take that tree down.

Well, under American law when we talk about, you know, public commenting periods or consultation with Indian tribes, all the government has to do is give these communities a chance to explain their concerns to them, but they do not have to change the project based on anything that's in this consultation.

They don't have to move the project if literally every single person they talk to tells them "Don't do this."

So it's very much just a situation where they have to give everyone a chance to express their concerns and then they can promptly dismiss those concerns and move ahead with the project.

"I'm Overwhelmed"

Will Falk: The Thacker Pass lithium mine project is a great example of government agencies knowing that they don't have to do anything about consultation, and government agencies that are often under a lot of pressure to permit these mines quickly cut consultation out. It's an easy corner to cut.

They don't like doing it because it's often confrontational. They are often forcing their project on top of a community, and communities don't like that.

So they often just try and get by on the smallest version of consultation that they possibly can.

For the Thacker Pass project, in particular, I think there's some sort of flagrant abuse of the word "consultation".

The Thacker Pass lithium mine was fast-tracked under a Trump-era secretarial order and that secretarial order required the local BLM field office (the Humboldt River field office) to conduct the whole National Environmental Policy Act process in less than a year.

Lithium Nevada Open Pit Lithium Mine (Construction), Thacker Pass, NV

As part of what we call the NEPA process, there are other laws that they have to take into account like the National Historic Preservation Act and the National Historic Preservation Act is what I've been helping the tribes litigate.

When you order a government agency to finish a permitting process in under a year, especially for very complex project like the Thacker Pass lithium mine project, you're putting those government agency employees under a tremendous amount of stress.

They have to consider all sorts of science; they have to document all the environmental harms that these projects are causing.

They have to consider the socio–economic harms that these projects would produce, and they, the Thacker Pass lithium mine project, put a lot of stress on the local Humboldt River field office.

Through litigation we've acquired all of their records, the emails between government employees where you can see there's several examples of government employees telling their superiors, "Look we can't get this done in a year it's just too much," "I'm overwhelmed," "I just can't do it."

And you know there's the email equivalent of a shrug and say, well that's your orders- you got to do it.

Maybe something that I'm more well versed on, or more able to speak directly to, is the way that the Bureau of Land Management misrepresented the extent of consultation that happened for the Thacker Pass lithium mine.

So if you go to any of the government's public-facing documents, documents like what they call the draft environmental impact statement, the final environmental impact statement, the record of decision and then other documents more specific to cultural resources like The Historic properties' treatment plan.

In those documents, BLM reported that they had been consulting with three tribes since 2017. Those three tribes were the Fort McDermitt Paiute and Shoshone tribe, the Winnemucca Indian Colony and the Summit Lake Paiute Tribe.

Lithium Nevada Workers, Thacker Pass, NV

Well, there's no documentation in the administrative record, which is all the documents that the government keeps and prepares, to prove that it complied with its obligations under the law–there's no record of this consultation in the administrative record ever happening.

And what's worse is the permitting process for the Thacker Pass mine happened throughout 2020, the tribes were particularly “hard hit” by COVID.

The law that's sort of the big, “800-pound gorilla” in these federal public lands projects is the National Environmental Policy Act.

The National Environmental Policy Act was passed and drafted in the late 60s as part of that whole wave of new environmental legislation that came along in the 60s.

Congress passed that law with the reasoning that environmental concerns weren't being taken into account for big federal projects.

The law is very careful to say that the National Environmental Policy Act process is just a review process; it is not something where if there is massive environmental harm, the project has to be canceled.

That's not what it says. It only says that federal agencies have to document those harms and provide that information to the public.

That's a really important point for people to understand because I think sometimes people think that when you hear about how destructive and how polluting a project is going to be, that in and of itself will be a "trump card" that shuts the project down.

That's not the case. It's laws like the National Environmental Policy Act just make government agencies tell the public about these harms before they put the projects in.

No Documentation

Will Falk:The National Environmental Policy Act (NEPA) is supposed to make the government agencies consider environmental effects, and as you said, effects to the "human environment" are part of that, and effects to the human environment involve things like effects to Native American sacred places, Native American cultural resources.

Then, a parallel law to the National Environmental Policy Act is a law called the National Historic Preservation Act. The courts have explained: what NEPA does for the environment, The National Historic Preservation Act does for cultural resources and historic properties.

The Thacker Pass project would destroy a literal modern-day Museum to Paiute Culture.

Paiute Elder Josephine Dick with Dorece Sam at Ox Sam Camp

One of the official legal terms for what Thacker Pass is, is a Traditional Cultural District, meaning it has a number of sacred places and sacred districts within the traditional Cultural District.

There were two massacres that happened in Thacker Pass. One of which took place September 12, 1865, carried out by members of the First Nevada Cavalry, a federal Cavalry Regiment raised during the Civil War.

This massacre was conducted as part of what's called the Snake War. The snake war was fought by Great Basin indigenous peoples against primarily mining interests from invading their land. So, in 1865, you have at least 31 and maybe as many as 70 Paiutes massacred.

Now in 2023, the government is going to destroy the evidence of that massacre. The very federal government that killed those Paiute people are now going to erase the evidence that it ever happened.

Robert Lundahl: Well, there are some aspects of this that are relative to the landscape, and maybe you can clue me in because there are two massacre sites.

My Paiute friends were saying well, you know they just dropped the bodies right there it wasn't like they stopped and dug a cemetery, so the bones sink right down into the Earth, so presumably, from their accounts, they're still there, plus whatever tools they had, and whatever living implements they had, were abandoned at the site of the one massacre.

Now, is that actually on the project APE or is that on the ranch next door? I keep hearing that they kind of shove it over here (conceptually), and not over there.

Will Falk: The way that the massacre happened, the first couple companies of the First Nevada Cavalry were patrolling the Quinn River Basin, which is the Basin to the East of Thacker Pass.

There are something like eight publicly available accounts of the Thacker Pass Massacre including a couple newspaper accounts that were published just a couple weeks after the massacre and that were based on soldiers' reports about what happened.

The Cavalry of course had lots of horses. In the Great Basin you have to stay around water sources with your horses, in such a dry place, and often the best grass forage was along the river.

So, they're going along this river, and they see campfires up in Thacker Pass and decide that they're going to attack.

They attacked the camp in the early morning hours of September 12, 1865, and the soldiers described a running battle that took place over several hours and over several miles of ground.

Thacker Pass, traditionally, for generations was known as a Paiute stronghold.

Nipple Rock, aka Sentinel Rock, Thacker Pass, NV

You can see the approaches coming from just about everywhere for miles. There's also a very distinct geographic feature that's been named "Sentinel Rock." I guess in Paiute they call it Nipple Rock because it's shaped like a nipple.

Even in the name Sentinel Rock, you get the lookout aspect– so they were camped below Sentinel Rock, and they would have fled up into Thacker pass.

All the Paiute accounts of the massacre described running up into the hills and caves and stone outcroppings that make up the north side of Thacker Pass.

If you go there and see it, it's a perfect place to hide if you need to, and it offers routes to other places that, if you're not from the Thacker Pass area, you probably wouldn't know about, so a great place for native people to run to, to escape.

An Astonishing Amount of Lead

Robert Lundahl: So, they didn't have horses?

Will Falk: They might have had horses with them, but they were surprised so they probably wouldn't have had time to jump on the horses. There's one account, the Ox Sam account.

Ox Sam, apparently, was able to get on one of the horses and take off, and the soldiers do describe a few people getting away on horses, but the vast majority of the people probably wouldn't have been able to get to the horses because they were surprised, and they didn't realize that they had the need to be concerned.

Even the soldiers' accounts describe this running battle that took place over several hours.

What's really interesting about how this works out legally, is that there's no mention of the Thacker Pass massacre in any of the permitting documents, so if you went to the final environmental impact statement and you wanted to read about cultural resources or historic sites, there's not going to be any mention of the September 12, 1865 Massacre.

This is a problem because the Bureau of Land Management possessed, in its own General Land Office records, survey reports from the late 1860s where a U.S Deputy surveyor described the September 12, 1865, massacre, that is, where he found remains of that massacre.

The Bureau of Land Management never produced these records, never told anybody about them, and now when we have pointed this out to the Bureau of Land Management, they have finally agreed that there should have been a discussion about that.

They're trying to limit where the massacre site happened to a small part of the project area in an attempt to say that they don't have to slow down construction of the mine, that the whole mine will not disturb the massacre site and this contradicts all the accounts of where the massacre happened.

Ox Sam Camp with Tipi

It also contradicts the way a massacre site actually functions within a living community. You don't go to the battlefield at Gettysburg and think that the only places that are sacred are the places, the exact places where each individual soldier fell.

In order to understand the battlefield at Gettysburg you need to understand the roads that the different units approach the battlefield from. You need to understand places like, where Robert E Lee was sitting on a horse, evaluating the field. For a battle like this, the same thing is true of a massacre site, especially when the accounts of this massacre site describe so many of the landscape features in Thacker Pass that will be destroyed by the mine.

So if you're going to take a Paiute child up to Thacker Pass and try to explain the massacre site, you would have to include all of the things like the Montana mountains on the north side of Thacker Pass, the approach that the soldiers took from the river, the likely routes that survivors took, you would have to talk about things like the soldiers saying that *they never made a search of the bodies*.

The reason they didn't do this comes from a newspaper article describing another massacre in the Paradise Valley, which is close to the Quinn River Basin, but that newspaper account, after describing a massacre, explained how often you would find dead Indians several miles from where a massacre happened and they said it's because these Indians, "can take an astonishing amount of lead before crawling off to die."

You would have to picture– okay let's say the camp was in one place–all the Paiute ancestors would have run in these different ways and you have to think about the way someone who just got shot several times would be trying to get away and where they would fall.

Max Wilbert, Author, Activist, atop Nipple Rock, Thacker Pass, NV

With the Thacker Pass lithium mine going on, you can go to the massacre site, and now you're just going to see this huge open pit.

You're going to see all the mining activities and, of course, you'll be forced to remember that 160 years ago it was exactly this kind of thing that those people were murdered for, back in 1865.

Robert Lundahl: What about the First Amendment of the Constitution? Does that come into play as a sacred site? Can you visit and practice your religion, or your worship or your prayers for ancestors, or any of the things that are important spiritually?

Will Falk: This has been tried, and the Supreme Court has come definitively down on this.

The fact or their interpretation of freedom of religion as saying that if the government was changing projects based off Native American spirituality, that would be akin to the government making laws that prioritize certain religions over others and that of course is not allowed under American law.

So, they use this really heinous logic to say, actually it would be violating the First Amendment for us to require projects to be moved because of Native American sacred sites, it would be infringing upon the First Amendment rights of, say, the corporations that have a right to mine there.For us to protect sacred sites in that way, there are ways for Native American sacred sites to be protected acts of Congress that's becoming more and more of a news item with places like Bears Ears.

But, you know, we have to understand that if we're going to be trying to get Congress to designate more and more of these sacred places, we're going to have a very powerful mining industry fighting that "tooth and nail" and I don't think we're going to win those battles very often in the courts or in through the political process.

Robert Lundahl: I've heard people say, "they can create a National Monument wherever they want to."

A Radical Analysis

Will Falk: It might be a little more complicated than that, but Congress does have a lot of power and it's something that we always should keep in mind as a tactic.

Dean Barlese, Elder, Pyramid Lake Paiute Tribe,
Ox Sam Camp

If we can create enough outrage over things–like the destruction of Thacker Pass and make this such a national and international news story and reflection on the genocide of Native peoples, then perhaps you know we can get some of these places protected by Congress.

I really subscribe to a radical analysis, meaning I try to get to the root of these problems and a lot of people talk about how we need to change the 1872 mining law for example, and I agree.

I wish the 1872 General Mining Law didn't exist, but when we're talking about using these political processes to do stuff like that, we have to understand that the American mining industry is a huge part of the American economy.

If we think that the mining industry and judges and politicians are just going to let us change the 1872 mining law and a huge part of the American economy, I think we're crazy.

I think that the American mining industry needs to not exist. I'm completely against mining and I'm also against the global economy.

So, to me the only way that we're going to survive the total ecological collapse that's happening right now is for the global economy to collapse.

Those in power, those who benefit from things like the 1872 mining law know exactly how that law benefits them, and I don't think we're going to persuade enough of Congress to change it, and if we did, there would definitely have to be trade-offs.

Power concedes nothing without a demand. Power is not going to concede the mining industry without being able to find profit somewhere else.

Robert Lundahl: ...well, what the Ninth Circuit Court of Appeals said in our case against Ivanpah (Solar Power Tower–Outside Las Vegas), which was shifted from failure to consult to First Amendment freedom of religion, they said, Well, what can we do to rectify this situation? Do you guys want a gate? You want a gate you can walk in there? I mean we're not going to tear down the whole thing because it's already built, right?

Which is logical, you know, in a certain manner of speaking, thinking.

Yeah, values driven you know, like you're saying, nothing happens without a reason or a benefactor, and so there were plenty of benefactors as to the status quo, and there are plenty of benefactors up north.

So we've run through a lot of laws, basically, in our conversation, we've just kind of gone (waves hand as if to brush them away). They don't have the power to engender a different kind of process where there's more perspective or different points of view on the table with real possibility of results from any kind of negotiation.

They're Not Negotiating

Will Falk: One of the very concise ways I try to say it is: consultation is not consent and we need to remember that.

We need to ask, "why is it that the federal government can come in and force a project over local communities," over the objections of local communities. That doesn't seem very democratic to me.

I would hope more important people understand this and start to understand the power that the government has to destroy your community.

Robert Lundahl: So here's another example, Imperial Valley lithium where they have new and untested technologies to retrieve lithium from geothermal brine from 11 existing geothermal plants that are built on or in proximity to the San Andreas Fault.

So in a vulnerable landscape, there's a lot of excitement and boosterism about getting a "twofer". Basically, we get our geothermal and we get our lithium brine too. It's generated a lot of excitement because this is a very poor community.

They've been suffering with airborne pollutants from the Salton Sea, from the pesticides and organic chemicals that are in the lake bed, and when the lake dries up, they blow out, so it's a real complicated group of factors but there's no clear technology in a sense, because it's never been done before.

It's complicated, so where do we go with projects like that? Where it's not certain that they can produce the results that they want, or that they're selling, like Ivanpah, which, as you recall, became a 40% natural gas project because it didn't it didn't meet contracts.

There were issues with cooling–it was originally supposed to be water cooled in the desert, and according to the physicists they had to redesign it for air cooling and they lost something like 10% of the power output, and they couldn't meet their contracts to San Diego Gas and Electric.

So just looking at it, just looking at it from a business standpoint, I mean, none of this is certain, right?

Will Falk: I think what's happening is corporations, the capitalists, are very good at co-opting fear and co-opting concern.

Climate change is real, and we absolutely have to cut carbon emissions. A lot of people that believe in what I call "so-called green energy" are really concerned and their hearts are really in the right place, but what's never questioned is this whole way of life– this whole consumptive way of life that's hugely energy intensive.

We never ask, "do we need to slow down consumption, do we need to learn to live with less?"

We just look to new technologies to try and keep this party going on a little bit longer.

And I think what a lot of people aren't understanding, like the example that you just used, is how tied up all of these so-called alternative energy processes, how much they depend upon fossil fuels.

The Thacker Pass lithium mine for example would burn 11,600 gallons of diesel on site every single day– it's going to produce over 150,000 tons of carbon dioxide equivalent emissions every single year.

Water Pipes, Thacker Pass, NV

It relies upon sulfuric acid to leach the lithium from the earth, and this sulfuric acid is dependent upon oil refineries.

The market for that comes from oil refineries, or where you go to buy the sulfur for this is oil refineries, so the whole idea to cut carbon emissions is that these alternative energies have to be truly alternative– they have to replace fossil fuels and your whole process depends upon fossil fuels.

It doesn't seem very likely that you're actually going to cut carbon emissions. You're not actually going to replace these fossil fuels, and I think the numbers really bear this out.

Yes, there are more alternative energy sources coming online, but our fossil fuel consumption is still rising. We really need to see what's going on here.

The corporations know how to make money from things like mining and industrial projects like these. They just switch the type of energy that they're producing from these projects and think that that's going to save the world.

But you can't destroy more of the natural world to save the natural world. You can't burn more fossil fuels to cut carbon emissions, and that's something that I think people really need to understand and really need to think about.

Nobody that's fighting the Thacker Pass lithium mine project denies climate change. We absolutely know that climate change has to be fought, but we just think that more industrialism is not going to solve our problems.

Systems Thinking

Robert Lundahl: Well, with the Mojave there was no accounting for the fact that long-rooted desert plants sink an awful lot of carbon; there were no numbers attached to that, all the way through the process.

Then they're shipping all this Chinese steel over to build transmission lines and other implementations of industrial designs, and there's no discussion of that either.

Then you've got all your terrestrial transportation, your gasoline and your diesel, as you mentioned, in your project construction activities and everything.

I mean when you start looking at it all, it seems like they're releasing an awful lot of carbon in a situation like that, and I imagine the same would be exactly true for Thacker Pass, so is this the failure here that as a body politic we're not looking at systems?

We're not looking at systems, what would you call it, systems thinking, or systems-oriented design?

Will Falk: I think that's definitely at the heart of the problem. One of the ways I've tried to explain it is, you know I wish when people looked at any sort of automobile, whether it's an electric or fossil fuel-based automobile, they would see the destruction that is reflected in that car.

If we're talking about just one part of an electric vehicle, the electric battery, lithium is a small part of that electric battery.

In turn, the lithium extraction that is involved in bringing electric cars to the public is just a small part of the extraction that is made possible by electric vehicles or that electric vehicles require to exist.

We have to see this. What goes into the steel manufacturing that is involved in electric cars? We have to see the mountaintop removal happening in Appalachia for that steel manufacturing. We have to look at the pollution that steel manufacturing produces.

Like in the Mississippi River delta, we have to look at how plastic manufacturing that goes into your car windows has created cancer alley in Louisiana–or to go back to the electric car battery, one of the rare earth metals that's essential to electric car batteries is cobalt.

People are now learning more and more that cobalt is primarily mined in the Democratic Republic of Congo and the human rights abuses in those mines– child labor is a huge part of those mines.

So bringing a mine like the Thacker Pass lithium mine into production is going to make the market larger for that cobalt, which is going to mean more human rights abuses in Congo, more children being killed in those mines.

It's not just each one of these mines, each one of these extractive processes is connected to dozens of other extractive processes that are destroying dozens of other places just like Thacker Pass.

That's one of the things that people sling at us all the time, like we're just a not in my backyard movement, and no. By stopping the Thacker Pass Mine, we might make it so a few kids in the Congo can live for a few more years.

It's truly all interconnected, and if we can find these choke points where we can keep some of this production limited, then we're actually creating a lot bigger change than it appears to be.

To your point, yes, systems thinking is totally missing and that's got to change.

Robert Lundahl: My (tribal) contacts were clear in letting me know that the Chairman was not supporting opposition to lithium mining at Fort McDermitt.

They were looking forward to having another election, getting another chairman, and so the tribe itself was split. The same is true with the Quechan people.

Preston Arrow-weed said,"well, you know our chairman is not really... our tribe is not really weighed in. This way we're not really clear where things stand, and everybody's looking out for the money and the grants and all those kinds of things, like all those grants to the Imperial Valley area which has suffered such poverty and environmental Injustice."

It's attractive to think that the circus is coming to town, and people are going to get a bunch of money.

Will Falk: These mining corporations are brilliant at taking advantage of tribal poverty.

They come in and they promise these jobs, most of which don't materialize; they do things like they'll build a tribe a new community center for a few million dollars, when you know they'll be taking out a few million dollars every couple weeks in profits from their mine.

For Thacker Pass they're initially (stating) it's supposed to be a 46-year mine and they've said publicly that they'll look to expand it.

So, this is fractions of pennies on the dollar for the wealth that they're going to rip out of Thacker Pass. But especially when we're talking about things like that 1865 massacre, we have to ask why are the tribes impoverished?

It's been a whole process of destroying local food systems, genocide killing off native people, forcing them onto reservations in a very fragile ecosystem like the one that exists in Northern Nevada in the Great Basin.

There were plenty of natural grasses and wild game. Paiute people lived great lives prior to Europeans coming and wrecking all the natural grasses, wrecking the water sources, building the mines, poisoning the water sources, and so it's hard for people to literally find what they need to live on the land in Northern Nevada.

And then, you know, the mines are directly contributing to the to the erosion of the land there, and the erosion of the land's ability to support humans there, but then they say well we can offer you this money which you can use to buy stuff that we've imported from somewhere else that we've destroyed.

So, you know this cycle has to stop. I really think people need to understand that the answer to tribal poverty is not destroying more land, that is only going to put the problem off.

These mining corporations take advantage of that poverty, and they have no interest in actually supporting the communities that they're involved in destroying.

And if they hear that, say, one Tribal Council supports a mine, that doesn't mean that all Native Americans support the mine. That doesn't mean that all the other regional tribes that descend from people who used, for example, Thacker Pass, support the mine too.

There's a lot of tokenizing that these mining corporations have gotten really good at. You find this is, in a lot of ways, how treaties were signed.

Robert Lundahl: Across the United States, you find a few friendly Native Americans in a community; you prop them up as the authority, and then you steal their land and say they gave you consent to do that, and yeah... they make an “x.” They make an “x” because at one time, not many could read in English.

To me, this is the same process playing out in 2023 that played out in 1815 that played out in 1780.

How interesting the mining law passed in 1872, and when was the date of the massacre? The massacre happened in 1865. Right- together in time or mindset or intention or history...

Will Falk: Well, and to make this go even deeper... so the 1872 mining law, part of the reason it was passed was to sell cheap mining leases to this new land acquired by the federal government.

But it was those mining leases that were necessary to pay off the federal war debt from the American Civil War, and then you can't offer cheap mining leases on land that you don't control.

You have to go in and destroy any militant resistance to your presence there before you can offer a mining interest protection.

Resource, Debt, and Money

Robert Lundahl: We have war debt, and then we change the law, or we create the law, which also happened all over the West, you know, if you don't like the law just change it, right, and then you have the massacre or the war against the inhabitants of the new land, so war breeds debt, breeds extraction, breeds war.

I mean if I were writing a master's thesis I'd probably go there, you know, it just seems cyclically, repeating.

Will Falk: I don't know how radical we want to get but...

Robert Lundahl: You know, that's not very radical. That's just like looking at dates and realizing there could possibly be a connection.

It's resource, it's debt, it's money, it's giving favors to your friends or whatever.

I made my first film on the restoration of the Elwha River. So here was a river that had "been working fine for a few thousand years," and then they decided to put a dam on it because they wanted power for one activity, for the Bremerton Navy Yard.

Again, it's the federal government. And they put this dam in there and they were required by the state to have fish passage, but they can't build a fish passage because they decide to hang it from a narrow canyon because it's cheaper, and it never goes to bedrock.

So, there's the washout threat and it affects the downstream and the tribe living there, and so there's a continual threat and there's no siren to warn tribal people.

And it cuts off the tribe from their food supply from 1915 all the way through the Depression. People can't get food and if they're found with a fish, they're taken to jail.

I mean, that took me a while to get my head around, like what if I were in the shoes of the people that I was talking to, or how did you guys work this out or these kinds of questions that a filmmaker asks.... but really, it's genocide.

Yeah, it's slam, bam, thank you, Batman. You're out, you're freaking out, and we're in.

Will Falk: Even the judge in our case kind of hinted like, why are your tribal clients so outraged right "now"? All that stuff happened in the past!

Graded Land, Lithium Nevada Open Pit Lithium Mine (Construction), Thacker Pass, NV

That genocide is ongoing– it's taken a different form.

The resistance to the Thacker Pass Mine has been non-violent and it will remain non-violent, but one of the reasons why it has to remain non-violent is we know what happens when you resist the federal government for things like mining.

The 31 Paiutes that were murdered in Thacker Pass are the perfect example of what happens.

Robert Lundahl: MRAPS, helicopters... yeah, when I talked to people (then) from Standing Rock, Lakota folks, they said no, you can't really talk about this, don't build it up, don't make it seem exciting.

Don't talk about the battles, the street battles you know, or the land battles, don't talk about that because we're under military occupation. For real.

Will Falk: I just really wish people could see that more clearly and understand that when you're advocating for things like electric car batteries, then you're also advocating for genocide, because you cannot have these products without genocide.

That's just how it is, and you can try and rationalize your way out of it, but I just don't think you can in good conscience.

I wonder if we should still call a lot of people environmentalists. I think they're more asking how "do we keep this way of life going a little bit longer?"

So, in some ways they're more like consumer advocates or they're apologists for this system. They might want to do it in a little bit kinder, gentler way, but I don't know what happened in the environmental movement where we thought it was okay to blow up mountains and call it green.

That's not an environmental movement I want to be part of. If we're going to withstand the collapse that we're going through we're going to have to be what I call biophilic or biocentric.

We're going to have to put the needs of the land first over the needs of wealthy people to buy electric cars, and that's a hill I'll die on.

Robert Lundahl: Yeah, it doesn't seem to matter whether it's copper, or lithium, gold, or uranium. There's no way to mine in a good way. You start digging up the Earth and that's not going to work without a more systemic understanding and practice.

I really thank you for laying it all out there. And I don't think you're a radical.

Will Falk: I wish more people, who I know love land truly, do love the land, who are afraid of what climate change will do, could broaden their perspective and realize that more mining is not the answer.

Chapter Four

Thacker Pass, An Archaeological Perspective

We want to do better, but we're doing the worst possible thing. And sometimes we confuse the two.

Imagine that the laws protect the most beautiful land, that is your home and home to living communities of animals, plants and other people who have been there forever, and someone wants all that to be graded down to dust, why?

Because... it's in the way of the wealthy, the greedy, and the well–intentioned.

And you hear, "I'm sorry, there is no time, let's move this along."

You may be number 3, or maybe you're number 10.

It doesn't matter.

It's go, go. go. We're moving.

–Bill Powers P.E., Who Are My People? Film, Robert Lundahl, 2015/2023

Austin Ringelstein joined the U.S. Navy before 911. After, he wound up on the bomb squad, defusing munitions in Iraq. Coming home, he applied his discipline and fortitude in the field of archaeology.

Now he's on a "bomb squad" of a different kind. He's helping to defuse the "bombs" of failure to consult and its resulting disharmony.

Austin addresses the "abysmal" attempts to come together in better understanding at Thacker Pass, and explains how a more harmonious balance with each other, and with the Earth, is possible.

Austin Ringelstein:

I ended up, the last three years, becoming the park archaeologist at the Santa Monica Mountains National Recreation Area, so I've sort of been thrown into what's called Section 106 of the National Historic Preservation Act, which is the main federal policy that guides archaeology and archaeological compliance, when it comes to federal undertakings.

A Reasonable and Good Faith Effort

Basically, whenever you have a federal undertaking, which could be any action that could result in impacts to archaeological resources or cultural resources, or any undertaking that is funded by the US government, you have to do an archaeological assessment to consider the impacts that the undertaking might have on cultural resources.

The key word there is that these agencies have to consider the impacts, which unfortunately doesn't give a lot of teeth to the law.

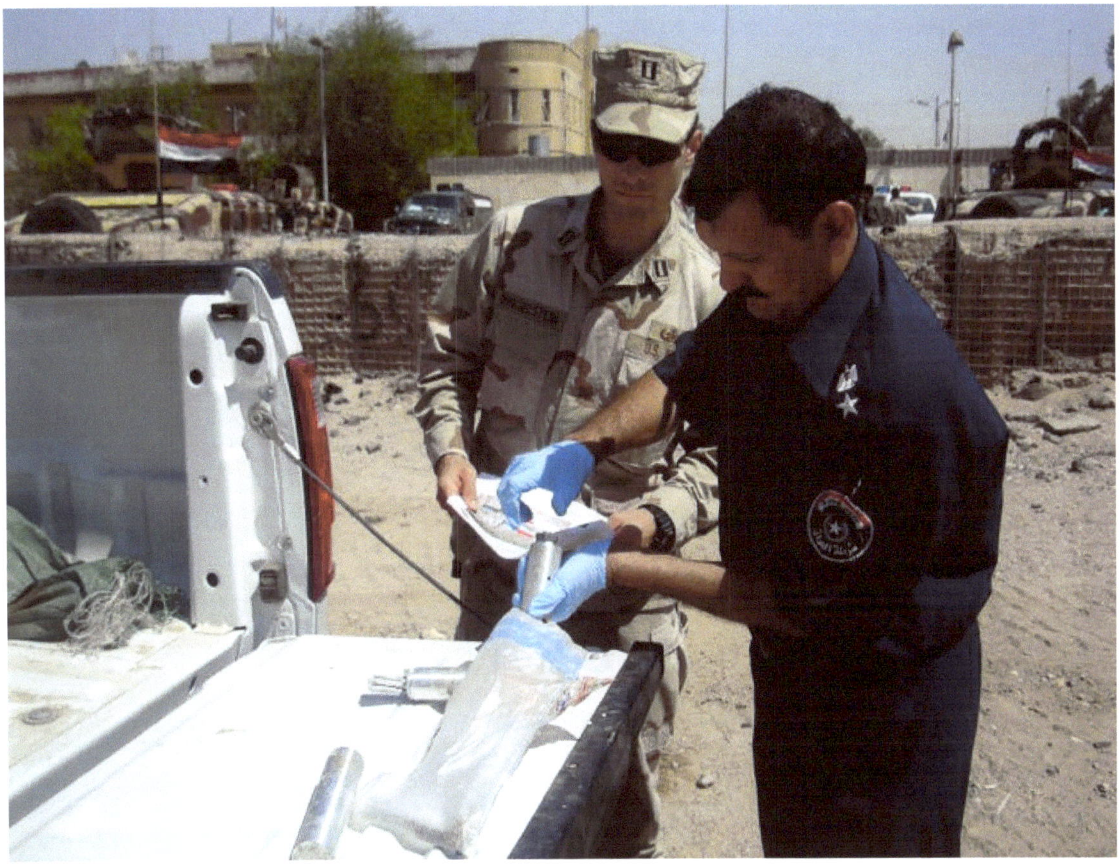

Austin Ringelstein assists an Iraqi police bomb squad commander to safely neutralize a bomb fuse in Iraq.

But it does at least mandate that an archaeological assessment is done.

It really has to be a good faith effort to survey an entire area, which is known as an area of potential effect, where the project might impact resources.

That reasonable and good faith effort is really all dependent on who's doing it, but luckily the consultation process allows for checks and balances to make sure that a good job is done.

That checks and balances process means that archaeologists are constantly working with the stakeholders, which in most cases are tribes, as well as the State Historic Preservation Offices.

So really, any time there's a federal undertaking you're going to do a pedestrian intensive survey of the area potential effect and then produce a report of your findings.

The last step is consulting with the stakeholders, as well as the State Historic Preservation Office to determine what the potential impacts of the undertaking might be.

Robert Lundahl:

Well, those are all processes that I'm a little familiar with and would expect, and as you say, the idea of a good faith effort, I guess, applies to what level of diligence, but also how much time they have, and how much pressure they're under.

As we were discussing, I was surprised by Thacker Pass, the quick permitting under the Mining Act, the fast tracking... Can you talk about that a little bit?

Fast-tracked

Austin Ringelstein:

I, of course, have worked in the Santa Monica Mountains National Recreation Area, but just having examined the EIS from Thacker Pass and reading some of the materials that have come out about Thacker Pass, there seems to be a consensus that the project was fast-tracked.

The Mercury News

10,000-year-old' Native American site found on...

From an archaeological perspective, you have to wonder- can you actually do a reasonably good faith effort when you have a condensed timeline? And when it comes especially to consultation, that's antithetical to that process.

The Department of the Interior has put out directives and policy regarding consultation, which say that you need to do consultation early. And often– it's more of an indigenous perspective than an existing capitalist perspective that we need to take time to build relationships and also to make sure that people can reach a consensus.

When you fast track a project, you don't allow that space for proper communication, for archaeological sites to be interpreted, for tribes to actually gather themselves and be able to speak about the value and significance of their resources, and to even to know what they're working with.

Really, the larger the scope of a project the greater amount of time should be allotted to that process. So, there's no doubt that fast-tracking a project because of a so-called deadline creates disharmony and is actually antithetical to the spirit of Section 106 and the cultural resources laws that we're guided by under federal law.

Robert Lundahl:

We're in the middle of this enormous push and pressure in industry and government for Green Energy Solutions and one of those is lithium for EV batteries for electric cars, the electrification of the transport system globally in response to climate change.

Then all that pressure is brought down. It seems like 18,000 acres is a lot but really you know when you're looking at a situation where we might say the whole world is watching, that 18,000 acres becomes extremely important, and the pressure gets extremely high.

An Abysmal Attempt

What do you think happened at Thacker Pass that was creating disharmony? And why is there a gap between the statements from Dorece Sam and others that you saw in the video and the process that they described?

For example, Dean Barlese says no one talked to us, no one came and talked to us.

And Dorece says it was a massacre site- so we know that they got surprised ...and all their living utensils and tools and hunting gear is out there, not to mention the bones that sink into the earth over time, not having been buried, nothing having been buried in two massacres.

I mean, it seems like an extreme situation.

Austin Ringelstein: When I look at it from an objective Section 106 standpoint, you have over 50 National Register eligible significant archaeological sites within the direct impact area, and then an additional 30 plus sites in the indirect impact area that are National Register eligible.

Austin Ringelstein

Well, in my opinion it does, and I would agree with that. Again, the larger the scope of a federal undertaking the greater the amount of consultation and due diligence to meet with the stakeholders and to come to some kind of consensus.

When you have 18,000 acres and over a thousand recorded archaeological sites this is a massive undertaking. So, the fact that three letters were sent to three tribal groups and there doesn't seem to be much evidence of follow-up to actually reach those folks, is in my opinion an abysmal attempt at consultation.

And really, the buzzwords here are meaningful consultation. Of course, that can be defined differently by different people but there has been policy and guidance put out that when you can, you want to meet with tribal members face to face.

You want to have really good listening sessions in terms of hearing the tribes, to be able to understand what their perspective is and understand what values they assign to different resources.

So saying that we sent letters out and didn't hear back during the pandemic and during a fast-track timeline is certainly not a good faith effort to have meaningful consultation.

I don't know the nuances or details of that process, but there hasn't been evidence to the contrary that there was meaningful consultation ever done.

The very fact that numerous tribal members, especially elders, have issues with this project in terms of their sacred sites and their archeology, and the fact that they have assigned values to Thacker Pass that are incredibly culturally significant, means that there is a conflict that hasn't had any kind of resolution.

There is evidence of an incredible amount of impact, adverse effect, that's going to happen to this place, and the bottom–line is that it was very minimally addressed in the EIS.

There hasn't been any viable pathway for mitigation spelled out in the EIS.

Appropriate to Scale

Robert Lundahl:

Well I think a lot of people in the public are unaware of all of this and when you look at it from their point of view, it's not unlike my point of view when I started the film "Who Are My People?", and I realized that no consultation had taken place.

That was difficult for me to understand, and it seemed wrong, unfair.

Austin Ringelstein:

In terms of your earlier question about these sites, in terms of fast-tracking the project and consultation and how that all played out in such a condensed timeline...

You know, one thing that I made note of was that while they may have potentially met some of the minimum standards for Section 106, they certainly failed to do meaningful consultation to assess the value of cultural resources to existing indigenous people, and the guidance by The Advisory Council for Historic Preservation, who should have an opportunity to comment about this project.

That consultation should be carried out in a manner appropriate to the scale of the undertaking and the scope of the federal involvement, so when you have a project this large, the consultation process should be done in a really thorough and thoughtful manner.

I mean, I've looked at projects that I've done with the National Park Service and our level of consultation was greater than this project for a small site and a very small undertaking, so it's actually quite stunning to me to see the lack of consultation and the lack of meaningful engagement with the tribes in this process.

The one thing that I really homed in on is that there seems to be a gap between sites that were identified as being significant, and the value of those sites to the living communities and cultural descendants.

My question is why isn't the BLM taking inventory of the value of the site to those living communities and cultural descendants?

Because, even built within Section 106, The National Historic Preservation Act significance, and the values that it may exhibit and the value to that tribe or to that stakeholder, really should be in line with one another.

Elder Josephine Dick and her niece, Dorece Sam,
Fort McDermitt Pai-Sho Tribe

So especially when it comes to burials, and we have a massacre site here, the special nature of burials is widely recognized in law and practice as having special qualities; they also possess a value to living groups that extend beyond the interest of archaeological research, archaeological assessment.

Why isn't there more time and effort being devoted to the value of this place in terms of it being a massacre site, and having people's ancestors literally, either buried there, or unfortunately scattered throughout that valley?

Robert Lundahl:

You know Will Falk, in the video we discussed, made the analogy to Gettysburg. And I've heard the phrase National Patrimony, so aside from this dichotomy of tribal versus non-tribal, awareness, and values assigned, from the standpoint of the country and what we learn and who we are and where our historical reckonings arise from, I thought that analogy was really good.

Also, his mapping, his discussion of the mapping of the site overall, the battle, it was systemic. Where did Robert E. Lee sit on his horse and watch the action? Where did the troops approach from, where did they fall- and then translating that into the Thacker Pass site, okay, you've got people running with a couple of bullets in them, where do they fall, those kinds of analyses... Can you talk about that a little bit?

A Historic District

Austin Ringelstein:

Yeah, actually Will did a really good job of spelling that out. What he's getting at there, in terms of cultural resource law, is that this is actually a designated historic district, and that means that in addition to the somewhat, not disparate, but separate archaeological sites that were recorded there's one cohesive unit that's been recorded as a historic district. That unit is, in itself, eligible to the National Register of Historic Places.

As a historic district, there's a holistic understanding of its significance.

Part of the problem with an undertaking within a historic district is that you can potentially adversely affect one part of the district which then destroys the Integrity of most or the entire place.

Will Falk, Attorney, Reno Sparks Indian Colony

It's been acknowledged in the environmental impact statement that the historic district would be adversely affected by these actions.

In layperson's terms, this is a major site or major location or locale for stone tool sourcing and stone tool processing.

Most of the over 1000 archaeological sites or cultural resources sites are what are called lithic scatters with obsidian tool stone production sites within the Thacker Pass component.

What that means is that most of the archaeological sites appear to be areas where people were doing stone tool processing, making arrowheads, making scrapers and other items that indigenous people were using for processing food, but they could also be ceremonial in nature. They could have multiple meanings.

It's important to point out also that there were grinding implements and other indicators of a broader range of activities that were taking place.

They were a little bit vague in describing that here, but I looked into it a little further and the grinding implements and the other indicators of a broader range of activities actually means that there is some habitation sites that they found.

I'm going to read what it says in terms of the proposed actions under alternative “A” effects in the mining area, and exploration plan area, would adversely affect all 56 historic properties eligible under Criterion “D”, due to the loss of integrity and data from ground disturbances altering artifact and feature assemblages, temporarily sensitive remains and the depositional environment of the local landform, and in the case of obsidian quarrying locales, connection to large-scale prehistoric land use patterns within the Thacker Pass component of the DHWOPD, which is the historic district.

Historic properties also eligible under Criterion “A”, specifically the CCC remnants would suffer the same integrity and data loss.

Robert Lundahl:

That's an interesting way of saying they're going to dig a 400 foot deep pit– I mean it's sort of like, we don't want to run the truck over that, kind of writing and really, you know they're just taking up a whole valley, about at least half of it, looks like.

Austin Ringelstein:

Yeah, I mean what I'm getting at here is that I'm not at all critical of these statements, because they're just true.

There's actually admission here that the project, that huge pit that you just referenced, is going to adversely affect 56 historic properties that are National Register eligible.

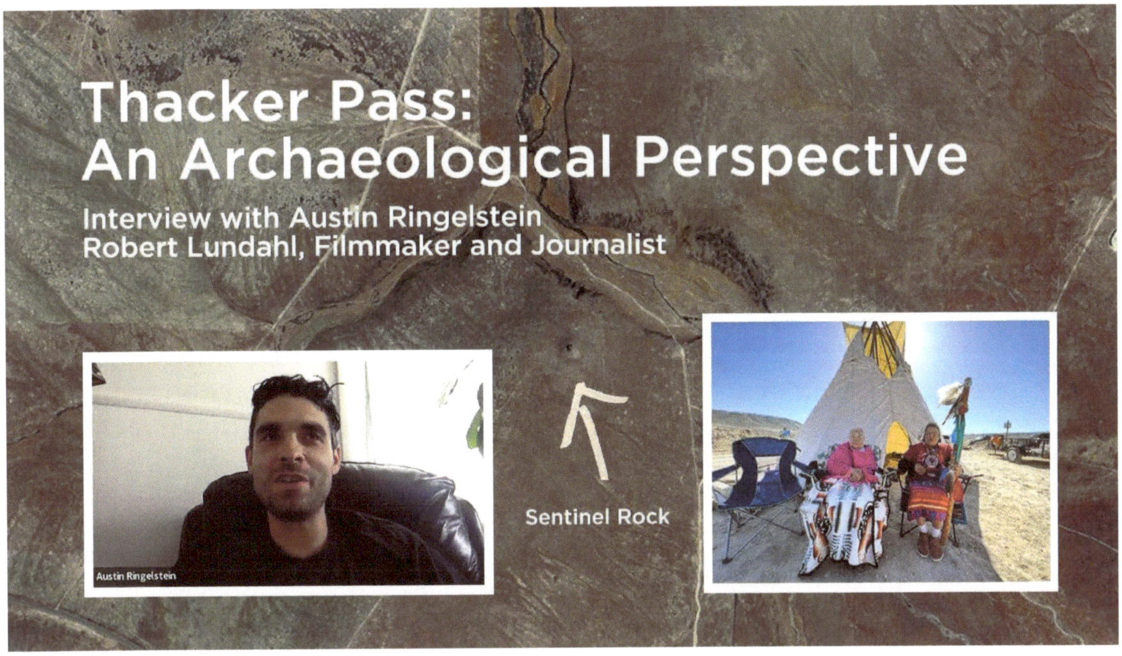

So the question here is, what are they going to do about it?

And the answer that they give is incredibly vague. The mitigation for effects from the proposed action is the implementation of an approved historic properties treatment plan, in other words an HPTP, which is currently in development.

So my question is, what exactly is going to be implemented to mitigate the effects on these 56 significant sites?

That's a question I can't answer without actually speaking with the archaeologists and BLM

Mitigo, mitigari, mitigavi, mitigatus

Robert Lundahl:

Well, what if there's no mitigation possible?

Austin Ringelstein:

Well, that's a great question. It's a really good question, Robert, because mitigation's already a tricky word to begin with. You know, I took Latin and it comes from the Latin mitigo mitigari mitigavi mitigatus, which means, in essence, to lessen or decrease so what you're getting at here is to lessen or decrease an impact, but the bottom line is there's still an impact- no matter what you offset or decrease.

I think, how many acres is it? Thousands of acres? You're having one of the largest impacts to an area that I can conceive of, and so mitigating something like that, it's almost inconceivable to understand how you mitigate it in archaeological terms.

What it might mean is monitoring and watching things either get destroyed or removed, and that's usually called a salvage effort.

But salvaging and mitigating are a little bit different because, how meaningful is your mitigation if you're just removing sensitive items and putting them in a different place holistically when it comes to an Archaeological District?

If you lose the setting and feel of that place that's a loss of integrity. There are seven aspects of Integrity that are analyzed in Section 106.

And within those seven aspects you have setting, feeling, those are just two of the seven, but those are the two that come straight to my mind that would actually be compromised by having an open pit mine there because you lose the original feeling in the original look, the original cultural value of a place that has incredible biota and incredible cultural resources.

Robert Lundahl:

Right, well it reminds me of when they built the Ivanpah solar power tower, and some of the statements that came from that time period– and I don't recall exactly whether they're attributed to Sierra Club or BLM, but in question was tortoise habitat and tortoise are protected, and this is an area that's primary tortoise habitat.

It's also an indigenous habitation site, and the statements that I heard were things like we're going to make more tortoise habitat that was a quote you know and it prompts jokes like I thought only God could make more tortoise habitat, things like that, but that's nonetheless what they said.

It meant moving tortoise, and when they moved tortoise at Fort Irwin Army Base down there in the desert they had a 50% kill rate.

So this idea of mitigation seems to be in the eye of the beholder. I mean, who benefits, and who wants what, and what you can say and get away with, am I right there or am I just cynical?

Austin Ringelstein:

Well, I think that, you know, working in the field I've worked in, mitigation is appropriate at certain times.

In the National Park Service, the bottom line is, we have trails and there are people that want to go out and enjoy trails, and they happen to sometimes intersect with archaeological sites.

It doesn't mean that you throw out the baby with the bathwater and get rid of all the trails, otherwise people actually wouldn't even have access to learn about cultural heritage or learn about the beauty of nature and resonate with all that's out there. So yeah, sometimes mitigation is totally appropriate.

You know, we're going to make this trail smaller, or make sure that when we do some vegetation trimming in this area that we do it with a monitor and that we do it in a really careful and thoughtful way, and that can be good mitigation.

But when it comes to a large-scale project like this and you're talking about mitigation offsets such as removing tortoises, desert tortoises to another area, or preserving one area so that you could destroy another, personally– and I know there's various opinions out there– I see that as problematic because it's another way of kicking the can down the road.

If you already have an area that's been assigned significance and value and we're talking over a thousand archaeological sites over 50 of which have been determined to be significant under Section 106, The National Historic Preservation Act, or as most people understand it, eligible to the National Register of Historic places, and we have a major area where we know a massacre occurred in 1865– which is significant, in my opinion, and those of many others, to the history of the United States, especially in terms of the history of the Indian Wars, the so-called Indian Wars, and in terms of the development of the West, and I'm not saying that that was necessarily a good thing, but something that we can learn from.

If we destroy these areas and then say we're mitigating it because we're going to create something else somewhere, somewhere else, then what are we really doing here? Are we really protecting our heritage, or are we just destroying one place and saving another until there's nothing left?

Yes...We Want Lithium that Bad

How do we reconcile kicking the can down the road when we're starting to run out of land to protect? I've seen this too often where we call it mitigation but really, we can't avoid the elephant in the room.

That there's actually destruction, there's actually cultural loss, and I think one thing that gets left out of these situations that we need to remember is that there's significant human emotion involved with these types of things.

So, you know, on paper, mitigation might look good at an agency level, but what do you say to the elders who are crying over the loss of their ancestors, and have to mourn maybe a second death because we want lithium that bad?

There's no denying that climate change is real and that we have an issue that we need to take care of in terms of the future generations and making sure that we can actually live on this planet.

But if we're doing that at the expense of current communities, hurting the communities that are in these lands where mineral extraction is taking place, then is that really doing our due diligence to help out future generations or are we just creating destruction in another sense?

Robert Lundahl:

At Thacker Pass we're talking to direct descendants of the last survivor of the second massacre, so their connection via family and via family stories and via how things worked and how people lived is really strong.

It's not necessarily dead, when they're talking about the presence of food sources on the land, and the presence of medicinal herbs on the land, and then the connections there with ceremony and religious ceremony, and prayer and practice, what I was hearing is they'd like to continue the use of that land and the way that it's always been used by them and they can't.

There was that comment about we used to pray at this certain area by, they call it Sentinel Rock, and in Paiute language they actually call it Nipple Rock, because it looks like a nipple. There's an area there where they pray that's now gated and they can't get in.

That's a BLM decision. That's a federal decision. You're locked out. Was that appropriate and is that a good example of cultural loss in the present day to present-day peoples?

Austin Ringelstein:

Yeah, and I hope it's not a loss, a permanent loss, and I imagine that if this fight continues that there can be a victory in the sense of reclaiming these areas.

Behind the Paradigm Shifts

What I'm observing is that federal law, especially cultural resource law, is behind the paradigm shifts that have happened in archeology and academia when it comes to honoring indigenous people and collaborating with indigenous communities.

The paradigm shift that's happened in archeology is recognizing that we're not just doing archeology to only examine the past or interrogate the past, but we're doing it to call attention to what's in the present and connecting the past present and future.

There's numerous, you know– I'm not the first to to say this– there's numerous indigenous scholars who are really leading that effort such as Dr. Tsim Schneider at U.C. Santa Cruz, and Dr. Sonya Atalay, in Massachusetts, who's Ojibwe.

Archeology can be a wonderful tool for actually vivifying, helping to revive and renew these relationships with the Earth and reconnecting with these places.

So at Thacker Pass, I think that what we see here is an archaeological process happening that's beholden to an outdated system and an outdated process that's constrained by time and constrained by the motives of capitalism to put profits and extraction over the needs of the people, and over the wishes of the people, and over the desires to reconnect with their ancestors and their spirituality.

And maybe it's not even just reconnecting, there's a continuity there.

There's a really great subfield in archeology called the archeologies of persistence and it's showing that a lot of these places and objects and sacred items... whereas, in the past they were they were determined to be remnants of acculturation, like if you have a glass bead mixed in with shell beads and it's a sign that there was some kind of cultural loss, and that it was less indigenous or less Native American than it was before.

But, on the contrary- what a glass bead can show is actually continuity and persistence because it shows that people were adopting non-local materials into their existing traditions.

I imagine that at Thacker Pass you have an incredible amount of continuity of practices, people continuing to gather there, people continuing to pray there over the centuries, and that story needs to be told and accounted for as part of its cultural significance.

I think it would be beneficial to the BLM and other agencies for these stories to be told. Why not make Thacker Pass not just a Cultural District but maybe a cultural historic site like Sand Creek?

Now that's not something I would determine, that would be determined through consensus. The tribes would be the ones to determine that but why not think about the infinite possibilities of the historically and culturally significant Place rather than thinking of it as an object of abstraction.

Beholden

Robert Lundahl:

Something in the way, something in the way of a grander ambition.

I was going to point out what we could call hypocrisy, societally, but I'd rather use the word disconnect because I think the reason that it's so important to mine lithium in the eyes of not only industry but policy makers is so that we can live in better balance with our numbers and with our machines and with our carbon dioxide releases, and so we can live in better balance, very simply, with the Earth.

And yet here you have this historic district and this historic people still there, who would like to live in balance with the Earth, and they're all too happy to share their history and their values and their beliefs and their understanding of why this could be done differently or better, not at all.

As a citizen of the United States of America, I feel like I've got an angel on one shoulder and a devil on the other. We want to do better, but we're doing the worst possible thing. And sometimes we confuse the two.

Austin Ringelstein:

I think what you're getting at is that there's sort of this aspiration by our society to want to reach our goals in terms of climate change, and live in balance, and so-called harmony with the Earth, and here we are trying to attain that using this ends justifies means approach.

We're going to extract lithium so that we can reach this goal by 2050 of being carbon neutral, and yet we're not listening to the very people who who are indigenous to this land, who are in accord with the land, and who are in accord with these values–that promote balance and harmonious living–not in some stereotypical sense, but in a real sense.

These are indigenous people who have a better understanding of the land and its resources and its value spiritually, and holistically.

Chapter Five

What Do We Mean by Environmental Justice? A Legacy of Greenaction

More sacred sites lost. All in the name of Capitalism and Greed.

Life takes us on circuitous routes. Do we follow a straight line of integrity? Maybe, but life has a way of picking us up and planting us back right down where we left off. Full circle.

After many years practicing my craft of filmmaking for clients, I engaged in a contract with a cause, raising awareness of toxic and contaminated sites nearby where people lived.

This led me to get to know a network of activists, native elders and Tribes, attorneys and allies. It was a route of evidence, a route of truth and reconciliation, a route of empathy and battle, such that after a while, though victorious and inspired, I needed a break.

But when invited by a Native American Apache colleague, a working knowledge of environmental and American Indian Law brought me back in, realizing that in 2023, laws on the books are conspicuously and tactically ignored.

Bradley Angel was one of the key individuals in this network, whom I had met in earlier days. He is the connecting force between several indigenous and non-indigenous storytellers I gratefully interviewed for the podcast series "Climate Change is Here" on Apple Podcasts.

For over 35 years, Bradley has been a local and national leader in the environmental health and justice movement and has helped communities win some of the most significant victories in the history of that movement.

In 1990, Bradley helped bring together grassroots Indigenous leaders in the first Protecting Mother Earth/Toxic Threat to Indian Lands Conference which led to the formation of the Indigenous Environmental Network.

In 2008, Bradley was one of five people from around the world chosen as a recipient of 2008 Lannan Foundation's Cultural Freedom Award in recognition of his decades of work with hundreds of diverse communities and Native Nations impacted and threatened by pollution and injustice.

In 2014 Bradley was the recipient of the 2014 "Environmental Justice Angel Award" from the East Los Angeles Community Youth Center for his environmental justice work empowering and supporting youth and Latino communities.

Bradley played a lead role in the landmark settlement in 2016 of the Title VI Civil Rights complaint filed by Greenaction and El Pueblo of Kettleman City that successfully challenged racially discriminatory actions by state agencies in environmental decision-making.

Bradley Angel:
My name is Bradley Angel and I am the Director of Greenaction for Health and Environmental Justice. Greenaction is a multi-racial grassroots organization founded in late 1997 by grassroots leaders from urban, rural and indigenous communities that I had worked with for about 10 or 11 years prior to forming Greenaction.

We work on the wide range of pollution, health and justice issues affecting urban, rural, indigenous communities, primarily in the west and very occasionally as capacity allows a little further– but primarily across California, Arizona, Nevada, and Utah- our main battlegrounds.

Robert Lundahl:
You've been working in Utah for something like 20 years and I know you spend a great deal of time there, so, can you tell me a little bit more about White Mesa and what's going on?

Bradley Angel:
Yeah, I started working with communities in Utah around issues of existing and proposed hazardous waste disposal facilities back in the late 1980s, before Greenaction even existed.

That was quite an interesting experience working in a lot of pretty conservative, rural, primarily Mormon communities, not known as environmentally-friendly, you know, didn't have that reputation but, in fact, everybody is concerned about their health, and everybody is concerned about the air they breathe and the water they drink and the communities they live in, so those were some early epic battles.

A lot of my early work before Greenaction was with Greenpeace USA. I was what they call the Southwest Toxics Campaigner, and I helped them diversify their organization at the time with my colleagues across the country.

I helped Greenpeace, at the time, work more respectfully and effectively with communities and I'm glad to say I was even proud that I was able to help repair relations that the big, giant, International Environmental Group had with indigenous people.

However, that was somewhat short-lived, and in 1997, Greenpeace International took over Greenpeace in the U.S and just about extinguished, wiped out the toxics program, and the environmental justice and Indigenous lands programs.

I was senior staff. I could have stayed doing something else, but I quit and submitted my resignation.

And one of the main reasons was not only just abandoning communities, which I wanted no part of, but communities really wanted a group that they would lead, and they asked me to work with them.

Through that period and then the battle led by the Mojave tribe and other Lower Colorado River tribes, an epic battle that defeated a proposed nuclear waste dump at a place called Ward Valley in the Mojave Desert.

That fight culminated in a 113-day occupation that led to the victory, and again through that work, it just continued a lot of my work with indigenous communities and native nations, for about 20 years now.

Because I live part of the time in Southern Utah, many years ago I was introduced to the White Mesa Ute community and which is part of the Ute Mountain Ute tribe, and White Mesa is a beautiful a little Community up on a Mesa, White Mesa, near the recently designated Bear's Ears National Monument.

The Last Uranium Mill

Bradley Angel:
White Mesa is just one of the most beautiful places you can imagine, so back in the 70s, the government let a company build a uranium mill in this beautiful place.

They built it pretty much on the fence line of the White Mesa Ute community and literally on top of hundreds of ancient sites including ceremony sites and burials.

For decades now the people of the White Mesa Ute Community have been exposed to radioactive and toxic poison and desecration of their sacred sites and a threat to their health, environment, lives, and culture.

So, I've been working with the White Mesa residents who have an organization called White Mesa concerned community. We're glad that working with them and increasingly the tribal government has been playing a pretty active and important role, really standing up for their tribe, for their community and environment and culture, and that is also just becoming an enormous battle of profound importance.

It's the last uranium mill in the United States, and it is turning into an international radioactive waste facility. So the battle is on and we could talk more about that.

Robert Lundahl:
Wow, well I've been to that area and I've been to Hovenweep. Do you know what that is?

Bradley Angel:
Of course.

Robert Lundahl:
That's just an amazing area. The ancient architecture and the way you can learn about how people lived by what they built, and a lot of it is there, and it's just a precious resource, a knowledge base for us and obviously very important to many tribes, so when Bears Ears (National Monument) was formed nearby, several tribes came together, right, and lobbied.

Bradley Angel:
Yes, there's a Bear's Ears Intertribal Coalition because the Bear's Ears is a vast area. It's named after a formation that looks like bear's ears, and in the languages of the tribes in that region, it translates, I understand, I've been educated about, into something resembling the words "bear's ears", and it really stands out on the horizon.

From White Mesa you have Bears Ears really close by, and directly to the east is also Sleeping Ute Mountain, which is one of the sacred places of the Ute, as I've again been educated about.

Real people with kids and lives and families and hope

Bradley Angel:
White Mesa sits not just in a beautiful scenic place, but one of profound cultural importance as well.

Real people with kids and lives and families and hopes and dreams live there, and every day they deal with the knowledge that radioactive waste from around the world now is being dumped in their backyard, on top of their ancestors, on top of ceremonial kivas, and ancient structures as well.

There are threats to groundwater, which the tribal government's environmental department has played a really important role in starting to document and going back and forth with the company, "who caused this?"

And you know, part of the reason that Greenaction exists is that, and why community people from so many diverse communities formed Greenaction.

Again, from West Oakland, California and Richmond, California to farmworker communities in the San Joaquin Valley like Kettleman City, Madres of East Los Angeles, San Ysabel, the Mojave Elders of the Colorado River Indian Tribes and others.

The main reason that these groups decided to form an organization that would fight and stand up for justice and take action and not sell out is because the government doesn't protect the people, the government doesn't uphold justice.

You know we have laws in this country and they're not good enough, we have environmental laws that exist that are violated aren't good enough but even the ones we have are not in many cases adhered to.

We have religious freedom laws, protection of indigenous peoples rights to practice their religion and culture.

We have environmental justice policies including from the Biden Administration, you know, a lot of talk about Climate Justice and environmental justice from the Biden Administration.

Trump of course didn't even pretend to care about that but you know for a decade you know for so many years we hear this rhetoric but the reality on the ground is different.

Disproportionate Impact

Bradley Angel:
And this is very true with White Mesa. It is an atrocity. It's an outrage and yeah, I should mention we also have you know the sacred site protection laws.

There are also civil rights laws which state that our government, if you receive federal funds you are prohibited under the law, Title VI the U.S Civil Rights Act, from taking actions that have what they call a disproportionate impact on protected classes of people.

That includes indigenous people, so how is it okay that the U.S government themselves, and also how is it the US government allows the state of Utah to turn this tiny community into an international waste facility that desecrates sacred sites, interferes with people and harms people's culture and well-being and is a threat to public health and the environment? And that's why Greenaction exists.

So, we were invited and play the role that communities ask us to play.

And in White Mesa we're glad we've been there on the front lines with tribal members, and the Tribe and allies now for more than two decades. And we're seeing traction.

We're seeing more support from diverse allies, indigenous and non-indigenous groups, getting a lot of media attention finally.

But of course, the Biden Administration spews this rhetoric that they support environmental employment justice and Indigenous rights, you know, and they've done some good things, like designated Bears Ears.

That's a huge, important National Monument. It's great and they should be commended for that. But right next to Bears Ears they're allowing radioactive materials, so-called alternative feeds, and other radioactive material to be dumped at the uranium Mill.

Robert Lundahl:
Can you talk a little bit about Free Prior and Informed Consent versus the... alternative?

Bradley Angel:
We hear a lot about the environmental justice movement; talks about and native nations that, whether it's a community or a tribe, should have the right to prior informed consent before their homeland and areas and communities are dumped on and poisoned by pollution and, you know, industrial pollution and threats.

To my knowledge, nobody went and asked the Ute Mountain Ute tribe if you want a giant uranium mill or if you want this uranium mill to turn into a giant international waste facility, before it happened.

The tribe certainly did not give permission to desecrate the burial sites of their ancestors, to destroy ceremonial kivas, import radioactive waste on trucks that go right through the White Mesa Ute community, this small community and so that's a really important concept, but in reality it doesn't happen.

And White Mesa is not the only example (where) it doesn't happen.

Decades ago, again, at the start of Greenaction and work that I was engaged in, just for a few years prior to late '97 when Greenaction formed, as I mentioned earlier, working with the Fort Mohave, Quechan, Cocopah, Chemehuevi, Colorado River Indian tribes to fight the proposed Ward Valley nuclear waste dump. Again, off reservation, but clearly it is sacred and traditional land.

The Ward Valley Nuclear Dump

Bradley Angel:
As I mentioned earlier, when the federal police were being sent in with the nuclear company and the police were going to try to clear the area of the indigenous and environmental justice protesters who had been camped out at the site, the tribes and Greenaction many others called for defensive, peaceful defense of the site.

It turned into a 113-day standoff and occupation against the police, until the White House had to cave in.

During that struggle, the White House sent the head of the Bureau of Indian Affairs and top White House officials to meet with the tribal government and to demand that– the US government actually threatened the tribes.

I was the non-tribal member invited by the tribes to kind of observe. I was in the back of the room with the elders as all the tribal presidents, chairpeople and council members were meeting with the White House guys, and the White House threatened the tribes.

They said, pretty much quoting verbatim, we do not, we the U.S government does not fund tribes engaged in criminal activity against the U.S government.

You're occupying federal land supposedly illegally. Take down the barricades or you will lose all of your federal funding, and for native nations that's huge.

And there's Dead Silence in the room. And I'm in the back of them thinking, damn the US government, right? This is yet another crime that's going to go down in the history books and literally threatening education, health programs, housing programs, jobs... ...and the tribal leaders asked the feds to leave the room.

The White House guys to left the room, and I knew this was going to be the decision, uh you know the kind of would that nuclear dump happen and the vice president vice chairman of the Fort Mohave Indian Tribe, really an amazing man and a hero and respected by all the river tribes, Llewellyn Barrackman.

He stands up with his cane and says,

“You know when I was young they took me from my family, and they cut my Mojave hair and they beat me because I just spoke Mojave, and they tried to strip me of my culture and they intimidated us and humiliated us and then he raised us cane and said we cannot be humiliated again. The barricades remain.”

It was one of the most beautiful things of all time in my life as an activist and just human being and everybody cheered and they said, “Bring the feds back in the room.”

And it was this beautiful moment where the feds come in really cocky, and you know, arrogant, thinking they're going to get the surrender, and the tribal leaders tell them *the barricades will remain until the nuclear dump project is canceled and the federal government's done.*

Before they stormed out of the room, Preston Arrow-weed, a traditional leader from the Quechan Nation turned to the feds and said,

“Which one of you is going to shoot me? Because that's what it's going to take to get me off this, out of the occupation and stop defending our land.”

And these guys just ran out the door. And it was great. It led to the Great Victory.

I take a couple of minutes to talk about this because that battle and victory has inspired the Quechan and Preston Arrow-Weed to stand up against cyanide leach gold mining at Indian Pass, which was defeated a couple years ago and might come back, but score one for our side.

Greenaction was proud to stand with Preston and the tribal members and tribal government. They're fighting another proposed cyanide leach gold mine called Oro Cruz.

It's inspired the people from White Mesa and beyond, all these people have now been supporting each other and that's a role that Greenaction has helped play is networking and sharing these stories, you can actually win. Do not give up.

That's really important for the people from White Mesa, so every year the Fort Mojave Indian tribe in February sponsors a Victory celebration.

We're now in, next one this February will be in February 2024, will be the 26th anniversary and I'm glad to say the 25th was the biggest gathering since the occupation.

It's hundreds and hundreds of people. And people from White Mesa came and Preston was there and it's just very inspiring. So when the people from White Mesa return north to their homeland in southern Utah, they return more inspired with more allies and friends.

And now we're working with the White Mesa concerned community and the Ute Mountain Ute tribe and many other allies but led by the grassroots tribal group and tribal member group.

The Tribe organizing now is reaching out and everybody's invited to come to White Mesa, Saturday October 14th (2023) for what's become an annual spiritual walk in protest.

It's a five-mile walk. It kicks off with a rally at the community center and a five-mile walk/ procession to the uranium mill's front gate and Bears Ears is there, you know, right there.

Bringing People Together

Bradley Angel:

Also, directly to the east is Sleeping Ute Mountain. This is what is so important: just bringing people together and providing backup and support and solidarity and encouraging people not to give up because you can win.

Robert Lundahl:
Well that's amazing because what you've described there, White Mesa, is a high-pressure situation and Ward Valley, obviously, as well.

You have an international community looking to dispose of their waste, so there's multiple energy ministers and countries and policy makers and, you know, they're cooking up the story for the locals and they have things invested in this, and that all comes down on a tiny little community.

Bradley Angel:
Yeah, we were talking about the early days of my work in the late 80's and as I mentioned, those were issues we fought at the time, to Greenpeace's credit, and it was a lot of my work also.

There was an explosion in proposals; industry proposals to build incinerators to burn toxic waste and garbage waste, solid waste, garbage, and some medical waste proposals.

The west desert of Utah was targeted. That was some of the work I mentioned earlier, all across California dozens and dozens of proposals again from the San Francisco Bay Area, countless proposals in the central San Joaquin Valley and the Imperial Valley and in East L.A and South-Central Los Angeles.

A lot of this was driven by an industry and government strategy of targeting what they saw as the most gullible, vulnerable, and easy to build.

You know, get through public opposition and this was easily the reality. You never see these things proposed in rich white communities.

In my work over the years I've dealt with hundreds of these projects and there's no doubt where industry and government targets.

The government had the nerve to put this in paper, so back in 1984 I believe, the state of California's Integrated Waste Management Board commissioned a consulting firm called Sorel and Associates and they did an Infamous study.

I'm just paraphrasing the name of the study, something like the political difficulties of siting waste to energy facilities, but basically it was a report, and it did not say go dump on poor people of color exactly, but it pretty much says that.

It also lists the characteristics of those communities they saw as most likely to resist, and we all know what those might be, and those least likely to resist, and those were rural, low income, in need of jobs, lower education, and religious. What we saw was primarily low-income people of color communities.

You also saw low-income and working class and highly religious and rural Mormon communities in Utah. And again, as I mentioned, hundreds of people from those communities rose up as well, although the state of Utah rammed a couple of those through.

But when industry ran into roadblocks and fierce opposition they never dreamed would happen from the little farmworker town of Kettleman City, or from the Madres of East Los Angeles, or black community groups in South Central, I'm sure it would have been great to hear those discussions.

It was probably damn! We've tried everything. Who's even more vulnerable? In their minds, where is it easier to get permits?

A Special Deal Just for Us

Bradley Angel:
In around 1988 - early ‘89 I started getting calls from indigenous people far across the U.S, saying, “can you tell me what an incinerator is? it doesn't sound good, but one's proposed for our reservation.”

I got a call for example from Navajo people, Diné people in Dilkon, Arizona on the Navajo Nation expressing what's an incinerator we kind of think it's not good. Can you help us? And there's a company called Waste Tech that says, “th*ey really like Navajo people, and they have a special deal just for us”*, and I was invited to come to Dilkon.

Actually that Victory led to the formation of the Indigenous Environmental Network, but lo and behold I then got a call in the same time period from Chickaloon Village in Alaska, "Hi, um, we were told to call you you might know something about this, um, we need help fighting an incinerator, there's this company called Waste Tech that says they have a special deal just for us".

Then I got a call from Vivian Jake and her colleague on the Kaibab Paiute reservation, about a special deal, "just" for that impoverished tribe, via a company called Waste Tech, same from Oklahoma. That's what industry was like in the 80's.

The beautiful thing is like 99 percent of those were defeated by the people and led by grassroots opposition in the tribes and sometimes by the tribal government, when they realized what was happening.

Targeting and dumping and dumping industry's waste, toxics and radioactive waste and stuff that would never be tolerated where the heads of government live, where the heads of these companies live, it continues to be dumped in places like the White Mesa, next to the White Mesa Ute community.

Robert Lundahl:
Well, we happen to be talking about quite a few of them these days, Thacker Pass, Peehee Mu'huh, in the Paiute language, and we're also talking about the so-called "Lithium Valley," which is in Imperial Valley.

When you're describing low-income communities and people of color and all their vulnerabilities, and we know their air quality problems because of the receding of the waters of the Salton Sea, and the blowing of airborne materials and dust containing herbicides and pesticides, and probably a lot of other chemicals that I'm not even aware of.

So they all have this in common, this targeting and pressure. Is there anything you can say about Lithium Valley? Because that's so obtuse in a way, you know the technology is not really being discussed because they say it's intellectual property and it involves removing lithium from geothermal wells in a second stage process.

No one seems to know how much lithium there is, how long it would be productive for, what the exact impacts are to water and air, and still I presume there are some in CO2 releasing and so forth from the construction and all of that, so there seem to be a lot of hidden costs with "we have a special deal just for you."

A Gigantic Public Relations Bonanza

Bradley Angel:
Exactly. And it's also the lithium boom. It is a very challenging issue, I admit, because it's being framed as a part of the Green Revolution, and getting away from fossil fuels, and to some extent, that's true, and the Thacker Pass project is very different even technology–wise than the Salton Sea and Lithium Valley.

Thacker Pass is just traditional- destroy everything, right? It's on an area very sacred to the indigenous people, a massacre site, and again, even though there's laws, some of which I mentioned earlier, oh it's no big deal, well let's see one of these things proposed in a mainstream religion's cemetery, right? It's just not going to happen.

In the Imperial Valley, it's really a gigantic public relations bonanza.

You have these companies joined by local elected officials and the state, all proclaiming this is the greatest thing since sliced bread.

It does appear to be different technology, but the issues that you've mentioned in terms of air pollution, impact on water, hazardous waste disposal is another one, are all issues that need to be looked at more.

The good news is they wouldn't be doing the surface, you know, destroying the whole surface of the environment there, but this is the Imperial Valley. It is already highly impacted.

It's primarily Latino but a lot of indigenous people native nations are there too, and we hear from some of our indigenous colleagues of concerns about the sacredness of the area, and concerns about environmental impacts such as earthquakes that could potentially cause damage from these operations.

When final decisions are made, I am not confident based on the track record of our government that they're going to properly consider what the impacts are.

Real expected impacts not make believe, real expected impacts from these operations, and what does that mean? What effect will it have on the large population of residents who already suffer high rates of asthma and cancer?

There, people work in the pesticide laden fields–there's terrible air quality– just all types of pollution issues there, and other social socioeconomic impacts that make people more vulnerable and the government's answer right now is all this is dollar signs, right?

But like we see so often, industry tries to placate– sometimes they don't even care, they don't even bother with that– but they try to placate in many instances including what we're seeing in the Imperial Valley opposition or potential opposition, with “oh well, we're going to donate so much, there'll be jobs and community benefits and things like that.

You know, community benefits are great if it's safe, but if it's not safe you can't put a price tag on people's health and lives.

Could drilling activities produce an 8.0 earthquake in the state of California?

Robert Lundahl:
Well, you know, community safety is a really interesting question. We were discussing this in one of my other projects, how to build a successful village of the future in Alaska, where many villages are suffering from melting permafrost and may even have to move.

We were coming up with these solutions of how to network people like you're saying, and the importance of that, and my colleague asked, “How can we translate this to Los Angeles?”

PHOTO SERIES : LUNDAHL, BRAWLEY ROAD TRIP IMPERIAL VALLEY, LEICA M8

And it wasn't exactly the follow-up question that I anticipated, but really it gets to the heart of what you're seeing. Here in California, we live in a larger community that includes let's just say for the sake of conversation, everybody, and in the Brawley geothermal field, apparently there was research done correlating the time and location of drilling activities with earthquake swarms in the area.

I know there's a similar study going on involving Dr. Michael McKibben and others related to that location of the geothermal activity at the south end of Salton Sea and its proximity to the San Andreas fault.

Apparently, it's much closer, but I have not yet seen research that would indicate that an earthquake swarm triggered in Brawley would have any less impact on the potential fault slippage that has been suggested could well produce an 8.0 earthquake in the state of California.

Lithium Rush

Bradley Angel:
I'm not an expert in that but these are all issues that really need to be considered, but if you look at the government websites and industry websites and those Salton Sea, Imperial Valley, the Lithium Commission, it's all rah-rah!

Our our colleagues in the Imperial Valley, including Comité Civico Del Valle, whom we work with on a wide range of issues, pollution and health and justice issues, everybody's trying to get a handle on what are their potential impacts.

How do we understand- is this technology that's being proposed in the Salton Sea, is it really safe? Yes, it appears to be different from, say, the Thacker Pass project but that doesn't mean it's okay.

There's a lithium rush and it's like the gold rush and they're seeing dollar signs. Greenaction's position is if something can be done safely and is really needed, but done safely without harm to people, health, environment, sacred sites, that's a good thing.

We're not convinced that's the case with the ; Sea projects and only if things are safe and acceptable to the community, then they can talk about community benefits and negotiate that with the community.

Too often when there are dirty industrial projects, and I'll give you an example. In California there's two commercial hazardous Waste landfills. There used to be three including in Westmoreland, right by Brawley.

That dump, thank goodness, closed just a couple years ago, but all three are in Spanish-speaking farmworker communities.

Not a coincidence. All three were approved decades ago with racially discriminatory permit processes. The Kettleman and Buttonwillow facilities that are operating today are operating on expired permits, and again, permits issued years and years and years ago with straight up racism, English only processes with police intimidation of residents.

I'll give an example. Kettleman City itself is an unincorporated town and so Kings County benefits enormously from the tax base from the biggest hazardous waste landfill in the west, run by Waste Management, Chemical Waste Management, The Kettleman Hills Hazardous Waste Facility.

The people of Kettleman City, you know, for many years they had no sidewalks, they had no clean drinking water, they had no streetlights, no infrastructure, and it's still pretty much like that.

The only reason Kettleman City has clean drinking water now is thanks to a civil rights complaint and settlement that Greenaction and the community group reached with the state in 2016.

This is another example where the company will donate for school uniforms, or they give out little trinkets and Frisbees. Could you imagine giving a Chemical Waste Management trinket from a toxic waste company to kids in their schools?

I mean this is what happens. So, they dangle the dollars, right? And our government lets this happen, right? So rather than our government taking care of our lowest income communities they use these as dumping grounds, government-approved dumping grounds.

Lithium Mining Location, Imperial Valley

This is what we call environmental racism. This is what is unjust. And as I mentioned at the start of our discussion, today, the laws we have in this country aren't strong enough, but they're not even enforced, right?

At Greenaction, one of our main missions or I should say strategies, has been to shove the existence of state and federal civil rights laws down the throat of the government.

Years ago, several government agencies told me at one point, "We don't have to consider civil rights in our permit decisions" around a toxic dumper, and some other nasty polluting industry. I said, "Oh really? I don't recall seeing an exemption for you in the law."

To Hell with Your Grants

Bradley Angel:
So we've been effectively using the law, including to bring clean drinking water to Kettleman City. But these are all issues that affect White Mesa.

White Mesa is a civil rights issue, an indigenous rights issue, a health issue, a justice issue. How is it that the Biden Administration, like every Democratic administration and I might even talk about the Trumps, but you know, because we knew which side they're on, they didn't even pretend to care, but the Biden Administration, Obama, Clinton, they all pretended to care about this stuff.

You know, Clinton had the executive order on environmental justice, but they don't follow these things. It's meaningless. Maybe they'll throw some grants out to some non-profits and that buys off some groups, and that buys their silence; it doesn't buy Greenaction off, that's for sure.

We're like, to hell with your grants to environmental justice communities and organizations, make the right decisions so we don't have to exist.

Do your job. But that doesn't happen in our society, where in our capitalist society where industry profit rules. And so, the battle continues, and we will continue working on the front lines with communities, with native nations, with our allies to protect health and promote justice, and achieve it.

A lot of things have changed in the thirty-five or six years I've been doing this work, but what hasn't changed is the anything goes for corporate profit attitude of both industry and government, and so this trouble for justice goes on.

Robert Lundahl:
Well, it sounds like some of these projects that you've discussed are under the Mining Act of 1872, not the disposal of the toxics and so on that we've discussed, but Thacker Pass, Lithium Valley, I think, wouldn't that be the case?

Bradley Angel:
I'm not sure. But again, there's outdated laws, right like the law you just mentioned is quite outdated. But even if certain laws, even if outdated, are on the books, there are other laws that the government is supposed to take into consideration.

There are environmental review laws. In many states from California to the National level, there's attempts to gut the minimal environmental laws that exist in terms of, you know, we're always accused of using some of these laws to hold up important projects, which is a bunch of nonsense.

Just don't propose terrible projects that will have harmful impacts on people and our environment and other living things, right?

Even in San Francisco right now there's an attempt by the supposedly progressive, mainstream mayor, to get some members of the Board of Supervisors to weaken environmental review in San Francisco, and that's the same on the state level, it's the same on the federal level.

So it's a never-ending battle and one of the reasons that as I look back, and Greenaction engages, we don't do a lot of lobbying or legislative work, very minimal, the most important things, but this is why I think that we continue to participate in regulatory and permit processes.

Protecting People's Health and the Environment

Bradley Angel:
We do our homework. We're not scientists, but we know scientists, and we're not technical experts, but we know them, and so we do our homework, but I think one of the most important lessons is that some things change and some things don't and action is really, in many cases, what it takes to win.

That's what it took to win in Ward Valley; that's what it takes to win from East L.A and Kettleman City to a cleanup site we're helping Comité Civico Del Valle with in Brawley, where the state was just going to do a partial cleanup of the PureGro pesticide contamination site– not that it's all *they're doing a great job* but they're doing more than they initially planned.

And we really applaud the indigenous people up at Thacker Pass and want to continue supporting that, but the reality is that our government does not protect people or people's health and the environment, and does not promote justice, certainly for low-income and working class and people of color communities, and that's why standing up and doing homework and fighting for justice has to continue and has to be successful.

Chapter Six

Chasing the American Dream in Lithium Valley

Environmental justice means economic justice. It means jobs justice.

A few years ago, on a road trip to the Salton Sea and the town of Brawley, Calif., I made photographs there of the bright sun, the hazy skies, faded signs, and general '60s milieu that reminded me of my childhood.

Some parts of town had a vaguely uninhabited look, boarded buildings sucking the life out of the place.

The Salton Sea qualifies as an environmental disaster area. Dead fish piled high on the shorelines whisper to the breeze, "all is not well."

The Salton Sea drains the fields of Imperial County, fields laden with herbicides, pesticides, and fertilizers, mostly hydrocarbon based, and some, carcinogenic.

It is the sediment, saturated with chemical residue, that blows in the wind, making people sick.

It was a few years and many realizations later that Imperial County began to embrace a would-be savior. Lithium extraction had come to town.

Luis Olmedo

My name is Luis Olmedo, and I am the Executive Director of Comite Civico Del Valle. We're an organization based in the Imperial County.

Robert Lundahl: Well, let me ask you a question about Imperial County. As we were talking earlier, the Salton Sea has numerous issues with agricultural pesticides running into it, and other chemicals as well.

As the water has evaporated over the years, the sand and sediment that contains all those chemicals blows and becomes airborne so that was the cause of a lot of problems.

Can you describe the way things were a few years ago, how you guys were looking at things and how you prioritized the needs of the community?

Luis Olmedo: As an organization we consider ourselves an environmental justice organization– so we see it from that lens as well as the fact that we're an organization that was founded by Farm Workers in the Imperial Valley.

This has been our home for the last 35 years. I've been with the organization for 23 years, so I grew up in Imperial Valley.

I've seen the Imperial Valley from being out with my parents working the fields, and I would say just kind of going through life and the years, a lot of times we can be very involved in our own day-to-day and not necessarily seeing air pollution or toxics where we live and the conditions we live in. A lot of times we ignore all of that.

As someone who's lived here, when I came into Comite– this was around 2000, my whole world changed. Everything that was part of the landscape, that was normal, suddenly was no longer normal.

And so some of the early things that I got involved with were around the things that, that we, as the blinders went off, you're like, "Oh I didn't realize that agricultural burning was a big deal, we see it all the time, we smell it all the time."

And it was actually people from, good friends from State Departments of Public Health or the Cal EPA or the US EPA, you know, as our network was already there. When I got here a lot of them started kind of pointing out things that I was unaware of.

Burning Season

Luis Olmedo: I still remember sitting down with some folks who worked at the State at the time, and then it was Burning Season, and they saw a whole bunch of mushroom clouds with smoke and, it's like, "What's going on? Is everything okay? What are all these fires?"

And I just looked back and I was like, "It's ag burning, what are you talking about?"

Oh, but they were so alarmed, and it was a real reality check for me when I saw their faces and their reaction to it. We realized how much we've normalized it.

Since then, as we took a deeper dive into our environment, we realized we don't have the best water quality, we're living next to contaminated and sometimes even superfund sites that we didn't even know were in our own neighborhood.

Air pollution, again, became a part of our language, our mission, and we got so involved that we started going out there and connecting to others.

We weren't born with the knowledge; we import a lot of knowledge too because there's other communities that have done this already.

So, we in environmental justice, we try to exchange ideas and best practices. We started going out there measuring the air and we started measuring the trucks, how many trucks, how many axles.

And we started measuring different parts of the community like, which is the cleaner community? Which is the most impacted community?

We started getting into understanding the Clean Air Act, and we got it to a point where we realized that as we kept persistently failing to meet federal standards, these are protections that we as residents, citizens, people who live in this community, we have those protections. In 2010, we had to file a lawsuit against the federal government because they were failing to enforce and afford those protections

PHOTO SERIES : LUNDAHL, BRAWLEY ROAD TRIP, IMPERIAL VALLEY, LEICA M6

I've been here 23 years, and we've been able to build a good team that has brought that expertise and who are working either within our own organization or working contracted, and so we've really built a huge arsenal of expertise to what it is today.

And so just as we have that history of having to learn over these years, we also have the history of how poor development has happened.

Anytime we circumvent the law, or elected officials try to cut deals because their metric is jobs, their metric is taxes– well, our metric is holistic.

We want to make sure we have good jobs. We want to make sure we have community benefits, and in those jobs, we want labor agreements, we want to make sure that in scenarios as opportunities open up, like they are right now with lithium and energy, those minerals are public patrimony.

Profit Share

Luis Olmedo: We want to make sure that if we are putting in a critical ingredient that is this public patrimony, that we feel it is a very fair expectation, just like anyone in the business world would expect, if you're an investor we're investing minerals we want a profit share out of that.

PHOTO SERIES : LUNDAHL, BRAWLEY ROAD TRIP, IMPERIAL VALLEY, LEICA M8

We want to be able to heal all the bad transactions that have happened historically, and we want to diversify our economy so we're not subservient to one industry that then controls our policies, our politics, pretty much our livelihood to such a degree.

No one should have that kind of power. And so right now is a very critical moment to be able to diversify our economy, to create a better path forward. A hundred years of agriculture, I appreciate them.

I love agriculture– there's nothing to do with that, but we cannot be dependent on one single industry. We need to be able to create other economies here, within our own agriculture economy and not necessarily compete with one another. We have room for more.

Robert Lundahl: Can we jump forward to the Lithium Valley commission? I know that you're serving on that with other representatives. Can you give me a little snapshot of what that commission is and does and who's participating in that?

Luis Olmedo: The Lithium Valley Commission was a great idea thanks to our Assembly Member.

We were able to have an opportunity for this commission to carry out a mission, which is to put a report of how this industry could successfully move forward, establish itself, and which would address many aspects, you know, from the socioeconomics, the environmental, being able to extract these minerals in a sustainable way. We had a very diverse group of Commissioners, and the goal of the commission was met, which is to produce a report.

We submitted that report, and you know this in December, I think in November, December of 2022, so since then the commission has already been sort of sunseted and retired.

We made that choice. There's a lot of moving parts to establishing this industry from the manufacturing to the extraction, to the environmental, jobs, there's just so many different... and we as commissioners, I think we all had different reasons.

We felt that retiring it was the best option. I can tell you what, and I don't share this much but I'm happy to say as to what my reasons were, I felt that there just wasn't enough representation from the fence–line community.

We want to make sure that there was a strong prominent voice from this community, and the commission didn't provide that.

I'm very concerned that everybody's looking out for their interests that they represent and those interests were very large and expansive in many directions there were many visions.

We want to make sure that we don't repeat the history of other mining towns that have been– that have had sort of a very short lifespan. There's a lot of extraction, a lot of harm, a lot of benefit, but then the legacy of harm stays behind, and then when that industry is gone these towns die.

PHOTO SERIES: LUNDAHL, BRAWLEY ROAD TRIP, IMPERIAL VALLEY, LEICA M8

We want to make sure that we have more say, more participation in the visioning, in the structure at all levels from the economics, the jobs, the community benefits, and the profit sharing, the taxes, and the way that they get distributed. We have an opportunity of a lifetime here.

A lot of times when environmental justice is discussed there's always this feeling of, "oh you know they're here to kill projects," and that is the wrong characterization of environmental justice.

Our goal is to be able to create and to be able to facilitate, to the extent that we can, a well-balanced economy, a sustainable economy. We don't want too much in any direction.

We want to make sure that all parties and all stakeholders are represented, because environmental justice means economic justice. It means jobs justice.

It means community benefits justice, all these things that, and let me compare it in this way, everything that is good for well-established beautiful communities, and there are many in this country, beautiful communities where they have beautiful landscapes.

They have the best education. They have the best tourist attractions. They have the best jobs. They have everything.

That's the American Dream. And so we shouldn't be then characterized as like, "oh they're anti–this or they're anti–that." No, we're for everything.

We are not for hoarding dollars for one special interest, or eliminating opportunities and assigning harm to disadvantaged, low income, or disengaged communities that have barriers in participation.

We're not going to be okay with that because we wouldn't be okay with that for others who participate and have political influence and access.

We want the American Dream, and the American Dream is for everyone.

A Legacy of Harm

Luis Olmedo: This is the best-case scenario happening in the Imperial Valley, or otherwise known as lithium Valley, right? In other areas, these projects have moved forward, and I wasn't there to witness it but we've seen terrible outcomes, a legacy of harm.

Communities have been harmed that will never be made whole again, and land that will never be used, at least in our lifetime or many lifetimes into the future.

People have a right to drive from one community to another and use our freeways it doesn't mean that they can speed; it doesn't mean they can dump trash on on the side of the road; it doesn't mean they can create a bonfire in the middle of a freeway.

It's the same thing with these industries. There might be a document that says you have a right to go out and and extract these minerals but there are laws that say otherwise. I'm not even going to go as far as expect people in business to feel like they're human right?

Business is business. They're about the bottom line. I'm not going to say, well you got to have morals, no. There are laws, and as long as you comply to the laws, and if those laws don't go far enough, then we'll go and make sure we put more laws on the books; but at the very minimum you meet all the laws in the book, that's what we're expecting.

It's up to us to become smarter advocates, and we are.

We're bringing in more available resources. We're working to be able to hire more experts, scientists, researchers, attorneys, NEPA attorneys, what we call leveling the playing field. Industry has the deep pockets to hire all these experts to look out for its best interest.

Government has their experts. Well now, in this case, in this scenario right now, thanks to philanthropy, thanks to our governor, and thanks to the changing climate locally with our government, they seem to be more in tune with equity and justice.

They are and it's either because they really see that it's in our best interest, or they see the dollar signs attached to these goals, and metrics around equity and justice.

Whatever the case is, we're in a position of power like we've never been before, and my hope is that that we're able to create a better blueprint as to how to extract these minerals because one thing we want and one thing's for certain, we need to reduce greenhouse gases, carbon from our atmosphere, because we are in a self-destructive mode right now.

We want to be part of that solution and one, a major contributor is carbon emissions from vehicles from large freight trucks, trains, to passenger- you know, family-owned vehicles and if we can produce a transition or I will call it a transitional option until we find the technologies where we can take advantage of the wind or water or other minerals that are less harmful to extract...

We know that lithium is perhaps going to be far less destructive than fossil fuel, so that's where we're at right now. We're really trying to do this. If this is the option, we want to make sure that it's done right and I do have to say that this state, in this country, is doing the right thing.

But also assuring that Native tribes have a prominent voice in this conversation because you know, as environmental justice we have partnerships with tribes.

Robert Lundahl: Let's go to the devil in the details. I don't know specifically what all the devils and details are but I kind of have an idea if you point over here you might find some difficulties, so looking at the Lawrence Livermore lab assessment of extraction of lithium from brines, they have said it holds great promise economically and for communities, but the other side of the coin is quote, “many technical challenges remain.”

So basically, I know you have these geothermal wells at the south end. You've got 11 of them and they produce water from beneath the Earth's surface, hot water. It comes up through pipes, and it provides energy in the current form and then goes back down in the ground.

So, the plan that you have, or the plan that is put in front of you is that the brine comes out of the ground and it goes to a secondary processing facility, or technology where the lithium is extracted. Then the water goes back down in the ground.

So, questions that I've seen are, that we don't really know how much lithium we can bring up, is that correct?

Do you look at things like this in your planning, because this is untested technology right? It's never been done so what do we have... a year's worth of lithium, or two years, or five years?

Then if it starts to run thin, do we have a market incentive to say *let's drill deeper further, faster, over this fault, this seismic fault*?

Can you address those issues and how you look at that in the planning going forward, how you plan to approach those kinds of questions, because these are world-class technology questions right? I mean, Silicon Valley doesn't have the answers, and it's up to you guys ultimately, I think.

Luis Olmedo: Yes, definitely, we have a entire community of researchers who are very eager to get the questions and then go out and seek the funding to go and answer those questions. I think we have to be very diligent how we do this.

We have to make sure that the type of science that is used that is helping to solve those problems and respond to those questions in a way that is usable, and translatable, and implementable.

Studied to Death

Luis Olmedo: I'll give you an example. As I saw and see a lot of times, we refer to "it has been studied to death."

We don't want to study this new industry to death. There's a limited pool of dollars, right, and researchers are always going to have the pressures of publish or perish.

We must be very diligent and make sure that our communities aren't just being mined for information, but that we have a coordinated approach as to what are those questions that we need answers to.

You brought up some of them, you know– how long is this resource there and sustainable, how much quantity, these are questions that I can point to like, one entity that is looking at them that's the Lawrence Berkeley National Labs, I know they're working on that. I've seen the University of California Riverside eager to go out and try to answer some of these questions.

Again, because very few times we get opportunities in this corner of the of the country, of the state, we want to make sure that, also, our local academic institutions are helping answer some of those questions.

Why? Because it's not going to be the only time, we're going to need that type of capacity especially if we're looking to diversify in our economy.

We know that some of these minerals are more difficult to extract. We know that lithium, as far as we've heard from the industry, we've heard from researchers, it is a science, you know.

It's not like you're trying to separate gold or these hard metals from an ore, but it's like you take it from a liquid to a solid and then to a liquid and it's very interesting, but I think a lot of this information, at least from the industry side, much of it is being characterized as "intellectual property". Now that's where we're going to have some trouble, I think.

And the reason why is because the communities have a right to be informed. We need to be informed as to what chemicals are being utilized, and what kind of risks are there to release those chemicals into the air, into the water, into our land, that are going to have a harm into our environment, and our entire ecosystem including human health.

I'm very confident that the industry is making lots of progress and some of them are talking about maybe as early as two years that they're being ready to go you know, sort of full-scale market scale production.

Again, I think in some cases we're still trying to convince the federal government where they have the sort of the larger resources to be able to invest in this industry.

I know the state is certainly convinced. You don't have to convince the state because the state wants this. It's really about the federal government and the federal government a lot of times is making decisions not really looking at the opportunity here in our own community.

Waste Streams

Luis Olmedo: There are going to be waste streams that exist. Right now just, you have three classifications of waste.

One is it's either hazardous, it's below the hazardous level threshold, or sometimes depending on the mixture of minerals you have some toxic minerals.

You have some very toxic minerals, and you end up with radioactive material that at times you know they have to push the emergency button and get all the cleanup crews in to handle that hazardous waste.

Salton City

So, with the lithium extraction some of the information we're getting, what some of the experts are telling us is that... if, in an average day, there's like 20 tons of hazardous materials, through the lithium processing 200 tons of material is going to be toxic.

Part of what we're doing is we're advocating and finding ways to create standards that California has, that the United State has, that are managing the waste, cradle to cradle, to the highest extent possible. It would be the dream that 100% would be recyclable.

We have considerable concerns. This is important that we have considerable concerns; one of the reasons is because we are the perfect candidate to end up with the harm.

We're the perfect candidate because we have a community that is considerably more ill in comparison to the rest of the state. We have people with respiratory problems when when COVID hit, per capita, we had the largest amount of death and permanent harm.

We also have large unemployment, poverty, it's just all the worst possible conditions, and so we're the perfect candidate as to where companies can really take advantage.

I would say that even government can take real advantage of us because our expectations are less than let's say you go to Beverly Hills or you go to Palm Springs which is our neighbor to the west Northwest.

I've been out there, and they don't even... I've never even seen a fast-food place, you know where you drive through.

For that reason, we are taking extra precautions to make sure that we don't end up with the harm, and in doing so, we create policies, and we hold those policies to the higher standards, so that we end up with the best possible outcome.

Chapter Seven

Dorece Sam: SLAPP Suits and Dirty Deals

When you look at a gleaming new electric vehicle what do you see? We love our cars, that is not in question, and our new technologies- and lithium mining for batteries for these vehicles is where it all begins.

I looked up "What is Green Technology?" on Investopedia and found this:

What Is Green Tech?

Green tech refers to a type of technology that is considered environmentally friendly based on its production process or its supply chain. Green tech— an abbreviation of "green technology"— can also refer to clean energy production, the use of alternative fuels, and technologies that are less harmful to the environment than fossil fuels.

We began this series on Lithium mining and the mining of other metals needed to create and upgrade energy systems which are considered essential. But there are problems of policy, of the enforcement of laws, and of environmental inequality and landscape destruction.

In my office, on the wall, was a page titled, "questions to ask about any new technology." It seems we can never ask enough.

Dorece Sam: We started the sacred fire on the water line on May 11th, when we erected the Tipi. So, I was on the front line for a week until the officer showed up on May 19th, towards evening, when they gave me that Temporary Restraining Order (TRO).

Dorece Sam Antonio holds her eagle staff on May 17, 2023 at Newe Momokonee Nokotun or Ox Sam Camp in Thacker Pass. Photo by Jarrette Werk (Underscore News / Report for America)

I believe it was like a week after the Temporary Restraining Order when... I wasn't given the lawsuit, the SLAPP lawsuit, it was actually emailed to me so it was just the heads up from, I believe it was Max that sent it to me, and so I didn't know about it.

So we have this lawsuit on us with myself being the only descendant of Ox Sam that was stated in this SLAPP lawsuit, and there are other descendants, Dean Barlese and there was two other Natives, BC and also Bethany that were named in the lawsuit, and Max, and I think it was Paul Cienfuegos and PTP, Protect Thacker Pass.

To this day they have never served me the lawsuit or emailed me the lawsuit. The only lawsuit I've seen is a lawsuit that Max shared with us.

Robert Lundahl: So yeah, he got one, he got the actual copy. He was served and then on the lawsuit it mentions you, too, so he was calling you and letting you know, is that it? Got it.

Wow, that's just unbelievable that you've been on your land for about 10,000 years and then they come in and they think, we're going to issue this lawsuit and basically the point of these SLAPP suits is to shut people up, right?

Dorece Sam: Right, yeah.

Robert Lundahl: Did that surprise you? I mean it's a lot of money, right, that they're claiming in damages? Dorece Sam: I guess it does surprise me but then, it doesn't scare me though, because I don't think it's really going to stand.

Personally, I think that the only reason why they filed this lawsuit [was an attempt to scare us] and the charges in the lawsuit were civil conspiracy, nuisance, trespass, torturous interference, torturous interference with contractual relations with prospective economic advantages and unjust enrichment (laughs).

'SLAPP' was coined to recognize lawsuits filed to silence criticism. Judge J. Nicholas Colabella wrote in Gordon v. Marrone (N.Y. 1992), “Short of a gun to the head, a greater threat to First Amendment expression can scarcely be imagined.”

When they filed the TRO, first off they can't pinpoint one particular person right? So they said there were four of us. It was myself, Paul, Max, and Chuck.

We were in all of the TROs, where they stated that we were flying drones really close to them, that the workers feared for their lives, which was all false allegations and accusing us of taking diesel from their vehicles, from their equipment, and I'm like, "I don't even have a diesel vehicle."

Robert Lundahl: That's odd. Yeah, well I've seen all the drone footage and when the drone is close to the workers, they're waving and smiling right?

Dorece Sam: Right, correct. So, I said it didn't look like they were in fear for their lives. You know, actually I think they were enjoying the attention.

Robert Lundahl: That's pretty harsh, you know, and the reason I asked the question was because it seems like with these large companies, and particularly the mining companies, all around the world, you know, they're using these kinds of tactics to intimidate people, right?

So, what about Sentinel Rock? Your Sacred Area?

Dorece Sam: They say it's in the work zone, only because they were passing the water line right through the road that goes to Sentinel Rock. So, they've already laid down their pipeline, so I don't even know if it's still considered to be in their work zone.

A cousin of mine worked with a lot of archaeologists and they told her that we should still be able to have access to Sentinel Rock, but I don't want to hurt the lawsuit.

I don't want to hurt how far the attorneys have come in fighting against them and things like that, so I don't want to hurt that, so I've been keeping my distance from Sentinel Rock.

It is hard because we're so used to being up there all the time and offering prayers and things and, you know, just walking around being on the land itself and it's hard.

But you know it's something that I have to deal with until this is over and hopefully that's going to be soon.

Robert Lundahl: What do you think is going to happen from here?

Dorece Sam: Oh my gosh, I hope they're going to dismiss it. It's just a scare, just to keep us away from there, and you know it's also causing a division because they've taken pictures off of the internet, off of social media and stuff.

There are people in there where you know they clearly stated their names, but these people they're not into a lawsuit. It's not that I want them in the lawsuit, it's just that it does look like they're trying to cause a division there.

Robert Lundahl: So, this started- what, a couple of years ago? I mean the first lawsuit that the tribes filed, there were three or four tribes, right?

So, there was Burns Paiute, Summit Lake, Reno Sparks, and they objected to the mine early on the basis of lack of consultation and following procedures incorrectly right, do I have that right?

Dorece Sam: Yeah.

Robert Lundahl: And you said you got involved later?

Dorece Sam: Max and Will first came to Thacker Pass, 2020, like around November, December.

Robert Lundahl: The winter...

Dorece Sam: The winter, 2020, they were camping at Thacker Pass, already. I... we didn't know.

Then my niece notified us, told us to go check on these people that were camping there. We had no idea where this was, because she didn't really give a specific place she just said oh, they're up on the hill.

My aunt and I were just driving around, you know, trying to find them but we didn't know they were out at Thacker Pass until later. We called her and she told us where they were out there.

And so, on March 3rd (2021), they were having a meeting in the Orovada community, and they invited us.

Myself, my aunt, my mom, my brothers, my niece, and my two twins we went out to Orovada and that was the first time that we protested.

We were protesting outside the building. I mean I had no idea what we were protesting. I knew that the mine was going to be there, and I knew it was going to be a lithium mine. For myself, I was happy there was a mine coming because I thought, hey, you know this is jobs for the people close to home.

I didn't know what kind of mine it was. That was the thing. And that was the first time that we met Max and Will, so we got talking to them and they started telling us the effects, what it was going to cause and how it was going to affect things, and that was when I got serious about this.

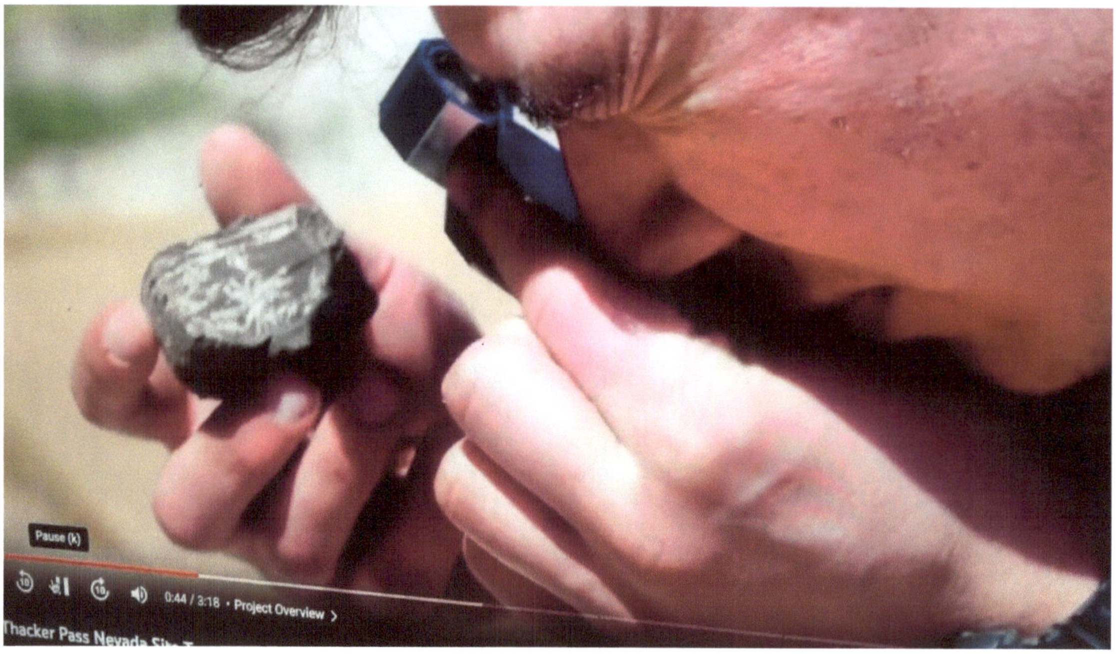

We knew that a Massacre had happened there and things, and so we've been involved ever since then.

And my son and I, I told him, let's just keep going out to Thacker Pass and let's keep droning and watch what they're doing and that's what we did.

You know we have seen from where it was all Sagebrush, just all nature and everything until now. Until you know, they have their brine ponds and everything filled up out there. I mean it's just they're going, they're not stopping for nothing. Robert Lundahl: Maybe they don't feel secure that they've really won those legal battles. I mean one of the problems that really blew my mind, honestly, was Judge Miranda Du, when she made that ruling that they didn't really complete the procedure.

She even went as far as saying the mine's illegal but we're going to go anyway- and it was illegal.

It's illegal because they don't have a permit to dump their waste, their tailings and their mess. They're going to dump it on public land, and they would need a permit for that.

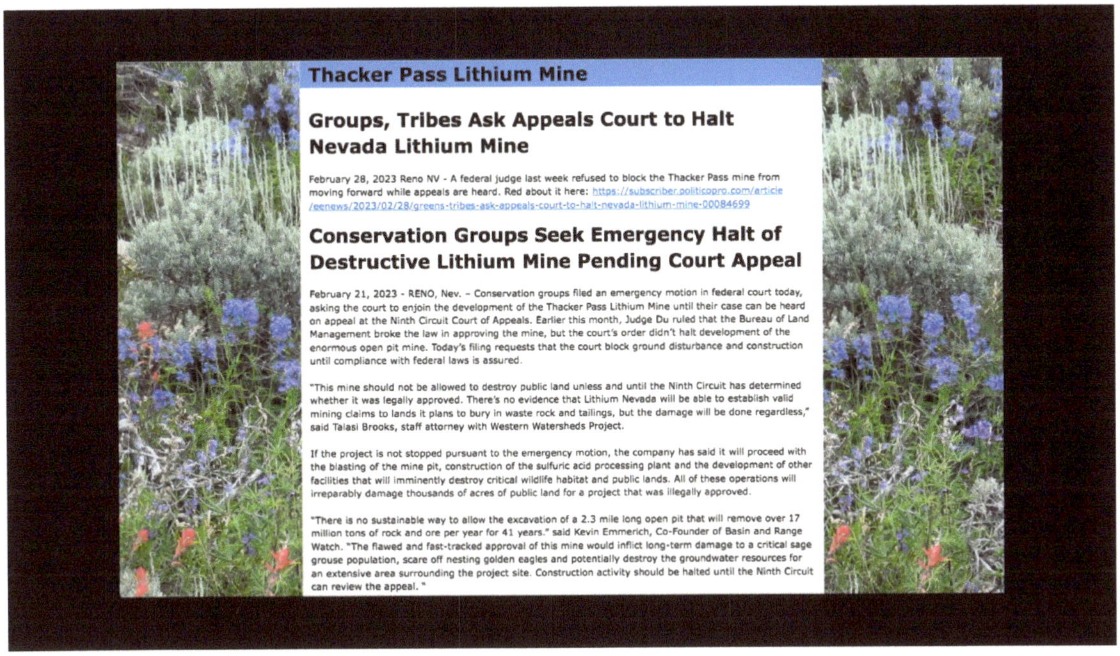
Thacker Pass Lithium Mine

Groups, Tribes Ask Appeals Court to Halt Nevada Lithium Mine

February 28, 2023 Reno NV - A federal judge last week refused to block the Thacker Pass mine from moving forward while appeals are heard. Red about it here: https://subscriber.politicopro.com/article/eenews/2023/02/28/greens-tribes-ask-appeals-court-to-halt-nevada-lithium-mine-00084699

Conservation Groups Seek Emergency Halt of Destructive Lithium Mine Pending Court Appeal

February 21, 2023 - RENO, Nev. – Conservation groups filed an emergency motion in federal court today, asking the court to enjoin the development of the Thacker Pass Lithium Mine until their case can be heard on appeal at the Ninth Circuit Court of Appeals. Earlier this month, Judge Du ruled that the Bureau of Land Management broke the law in approving the mine, but the court's order didn't halt development of the enormous open pit mine. Today's filing requests that the court block ground disturbance and construction until compliance with federal laws is assured.

"This mine should not be allowed to destroy public land unless and until the Ninth Circuit has determined whether it was legally approved. There's no evidence that Lithium Nevada will be able to establish valid mining claims to lands it plans to bury in waste rock and tailings, but the damage will be done regardless," said Talasi Brooks, staff attorney with Western Watersheds Project.

If the project is not stopped pursuant to the emergency motion, the company has said it will proceed with the blasting of the mine pit, construction of the sulfuric acid processing plant and the development of other facilities that will imminently destroy critical wildlife habitat and public lands. All of these operations will irreparably damage thousands of acres of public land for a project that was illegally approved.

"There is no sustainable way to allow the excavation of a 2.3 mile long open pit that will remove over 17 million tons of rock and ore per year for 41 years." said Kevin Emmerich, Co-Founder of Basin and Range Watch. "The flawed and fast-tracked approval of this mine would inflict long-term damage to a critical sage grouse population, scare off nesting golden eagles and potentially destroy the groundwater resources for an extensive area surrounding the project site. Construction activity should be halted until the Ninth Circuit can review the appeal. "

So, when the first lawsuit was filed, that was when they were saying that there's no massacre site, they had not admitted that. Then it was later, I don't know when, but it's like a year later or something they came around.

Dorece Sam: They also went to a meeting, their quarterly dialogue meeting, here on the reservation with Maria Anderson, she's one of the workers at Lithium.

So, they showed a video of the finds that they had out there, the arrowheads, and you know things like that, and they even showed a hearth, a kitchen that the Paiutes used a long time ago.

A hearth, and they had found two of them! They had taken a picture, and they were showing us where the fire had been around it, you know, the different colors, the layers of dirt from the hearth.

So, it's interesting that they have all this evidence and they show it, but still there's no stop to what's going on out there, you know the destruction, the desecration.

Robert Lundahl: Well, the hearth, I mean, that's like the family fire, right? I mean, doesn't that indicate that people were living there?

Dorece Sam: Yeah, they said it was like their kitchen stove.

Robert Lundahl: Kitchen stove, right. So that's a big admission. Wasn't there also tool manufacturing? The spear points, the arrowheads, wasn't that obsidian?

Dorece Sam: Yeah, they have all of that and they said they have them stored away in Henderson, Nevada, I believe.

Robert Lundahl: Okay so they admitted there's the village or there's people living there; two hearths, two families, two kitchens, that could be a bunch of people right there you know, depending on the size of the family group, right?

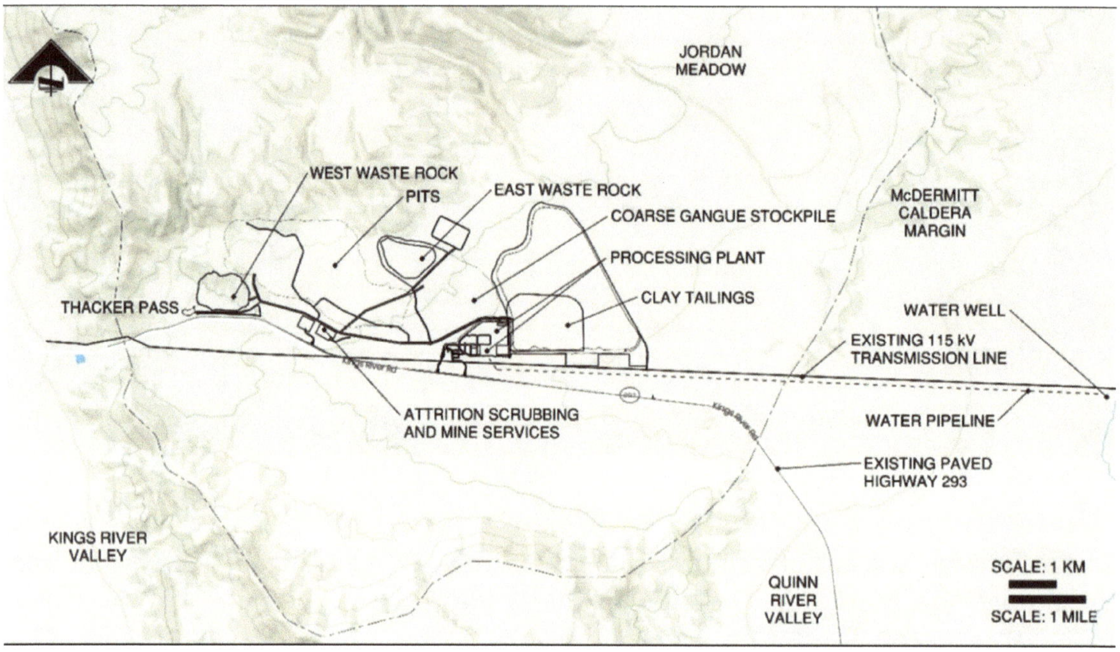

There's tool making, so they lived there, they hunted there, they had that all in proximity so that seems like a pretty good situation. They didn't have to go a long long way for the tools, for the spear points.

It sounds like a settled area, and that would explain the massacres because people we're living there, and then the Pit River people came in, that was massacre number one and they got pushed out from the colonizing, the gold rush, and then the Nevada Cavalry came in for number two, and people were living there and Ox Sam was there.

I just want to clear this up because there were two massacres, there's a village, there's people living, they're making tools, it was real settled down. It seemed like a good place. I mean people must have lived there well and easily and made a living.

Dorece Sam: Yeah, it's right by Crowley Creek, so there's like a creek that runs right there so it is in a good place. It feels really good there. I mean, I would have lived there– the fish, the deer, it's plentiful there and the herbs, they get the medicines, it's all there.

Robert Lundahl: We talked to Tracker (Gina) and she set us up with Austin Ringelstein, and he looked at the EIS, and he said, well, they really didn't do a comprehensive job.

They didn't walk the land, they didn't have time, they didn't follow through, they weren't particularly good at it. He called it "abysmal." Actually, "abysmal," because they're supposed to do all that in good faith and work with tribal people.

Dorece Sam: I didn't know.

Robert Lundahl: Then later we found out that Oak Flat (in Arizona) had that same situation where they they had been sort of railroaded into a land exchange and were set to turn the land over to a non-United States mining company and let them mine, but they didn't have a permit either down there to dispose of the materials and the tailings. Right.

President Biden put a pause on that, you know. He said maybe we're not going to go forward right now, that kind of thing, not that they couldn't do it later or approve it later, but they stopped it.

They didn't stop it when they did the same thing to you guys.

Dorece Sam: Well, how do we know the next president is going to be any better than the ones that we've had currently and previously? It's just like, you know, the situation here on the reservation.

Our previous Council chairwoman, she didn't communicate with the community. She didn't make phone calls, return phone calls, to our current tribal chairman.

He's for this mining all together, and it really makes me wonder because there's a few times where I've seen him going into the Kings River Community to purchase hay for his livestock and if any of the communities are going to be hit hard and first it would be the King's River area.

Makes me wonder why he is for this mine, you know. He doesn't think. He's not thinking.

Well, they set up our first hearing and then dismissed that to a later date, and I believe I might be wrong, they rescheduled twice already, and so I believe we're looking at somewhere in October.

That goes to show that you know they're just using this SLAPP lawsuit to keep us out of there, keep us off of our own lands.

The young people, they're very, you know, destructive. There's a lot of jealousy, a lot of animosity that goes on here. We had a nice Travel Plaza, and when our former chairman, Tilden, the one that fast tracked this mine, he got voted out, and someone went and set fire to the Travel Plaza.

They said that it was a grease fire that caused the fire there. Some say he had something to do with it, I don't want to say it was him, but I was told as much.

So, you know the Travel Plaza was a good thing to have there because some of the local Natives, you know, they sold their beadwork and "what not" there, and there were a lot of travelers that went through.

But that burnt down.

And you know, for myself the way that I see this culture center is, where are they going to get the things that are going to be there? They want a museum in there. But how will they get the things for the displays when the whole, you know, a majority of the community are against the Ox Sam Camp?

We don't want the mine, and they do want the mine. A lot of the family are into making buckskin moccasins and gloves and cradle boards, beaded items, and there's a lot of people who don't... don't know how to do things.

Robert Lundahl: And this is for the cultural center, so the building that they want to make a good public relations image is a culture center.

The BLM didn't acknowledge the presence of the people in the village and the massacre sites, and then a lawsuit was filed, and then they did acknowledge it and they even have photographs.

They brought them in and showed you the evidence that they had and they acknowledged it, which is like where you lived and how you cooked and obtained tools and all those kinds of things, so that was like really super clear.

And then you're worshiping at your sacred places where you usually have prayed, you know. "Usually" for 10,000 years... and then they called it their work area because there's a road going there and you're trying to go there too. ...and I want to talk to you about the excavator in a minute, but so then all of a sudden, it's their work area and then they sue you guys to shut everybody up.

Dorece Sam: Right.

Robert Lundahl: So, there's kind of a real evil process going on where they just don't care what they do, and they want to shut it down and you know attack the people.

Dorece Sam: Right, yeah.

Robert Lundahl: So okay, well let's talk about what happened with the excavator. When that young man went out there and talked to the driver, it was then they started talking about all this harassment stuff.

I saw the video, and he went out there, and he said, "What are you guys doing? Do you feel really good about tearing up Native land?"

You know, this is Native land, right, so I mean that was a very reasonable, intelligent conversation that wasn't threatening.

Dorece Sam: I just woke up that morning, and I see that picture of the young man running towards the excavator, and I probably know as much as everybody else does by the videos.

The young man, when he tried to stop the excavator, I guess was when the person driving the excavator was afraid that he was trying to hurt himself or that he was gonna get hurt, and so they called the officers in.

So, they kept calling the Sheriff's Department. The sheriff didn't want to get involved but being that their boss kept calling them and making them come out there.

I don't know if it was a planned thing because from what I was told, there was a couple of Sheriff's Vehicles out there so it might have been a planned thing that day that they raided the camp, and I think it was because of the excavator incident.

Robert Lundahl: So, when they raided the camp they took sacred objects too, right?

I mean, I heard about eagle feathers and that there were other sacred things taken.

Dorece Sam: Yeah, they took my prayer staff, and you know that staff held a lot of prayers; it had a lot of offerings and carried an eagle feather on it that was gifted to me from a friend of mine that lived in the Kings River Valley.

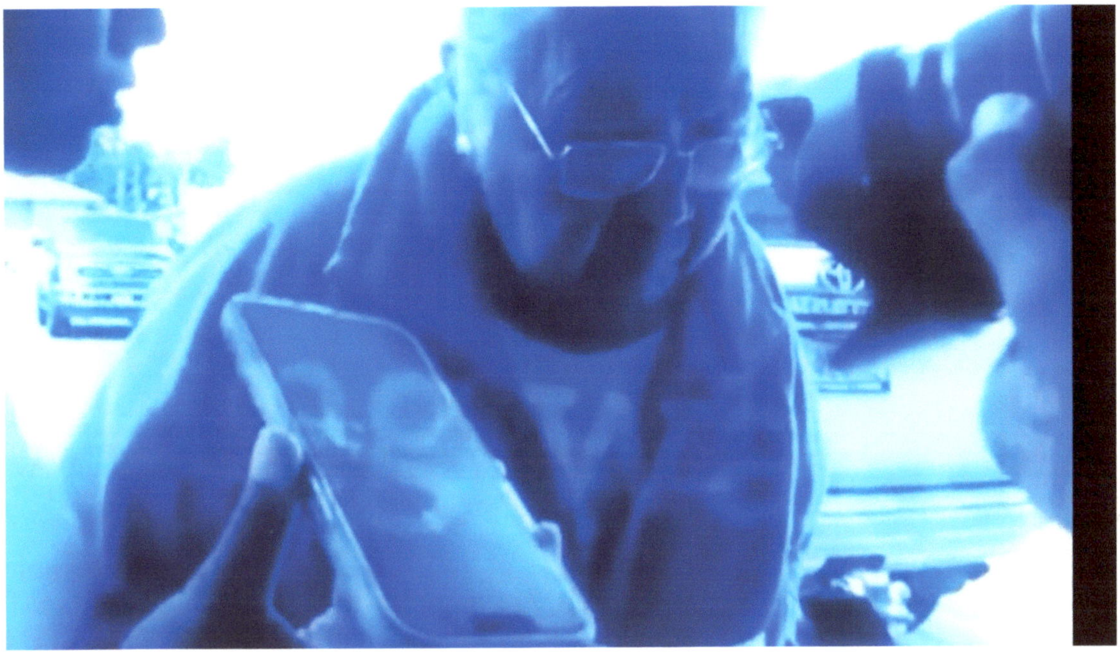

Retrieving prayer staff from police trash bag

He had gotten the feather off of one of the Eagles that nest right there in the Thacker Pass area. That is where that eagle feather came from, so that's why I carried it on the staff.

Robert Lundahl: ...and they gave all that back, right? But they put it in garbage bags and it didn't look like they took care of it.

Dorece Sam: That's correct.

Robert Lundahl: Josephine looked really upset.

Dorece Sam: I'm the president of the Native American Indian Church of the State of Nevada. And my Auntie, Josephine, she's the Vice President, and being that I couldn't communicate with Lithium or be in the Thacker Pass area, I told my Auntie I said, well what if we write up a letter for you and you can hand deliver it to the Lithium office for us to retrieve our items back.

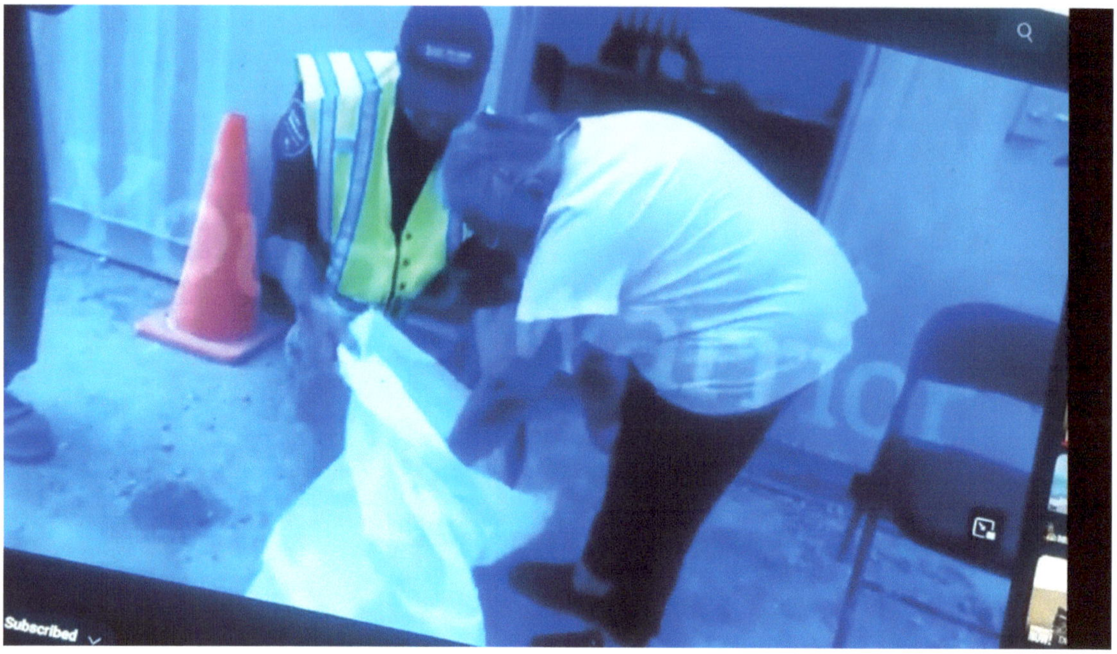

I told her, I said, I'm not too sure if they would, but if they (don't) you know they're violating a lot of things (laws), and so I started looking in my paperwork with them, because I told them they're violating the American Indian Religious Freedom Act (AIRFA).

We wrote this letter for my auntie, and she delivered it to... she tried to take it to the Sheriff's Office, and she just got the runaround, literally, all over Winnemucca.

And then she took it to the courts, and then she was told to take it to the Lithium (Americas) office, and she took it to the Lithium office, and I think it was there that she was informed that... the security at Thacker Pass had everything.

So, she and you know the rest of the crew went up there to retrieve all of these items and I guess her main concern was my staff because of the eagle feather on there.

And it hurt her to see my Eagle Staff, my prayer staff placed upside down in a garbage bag.

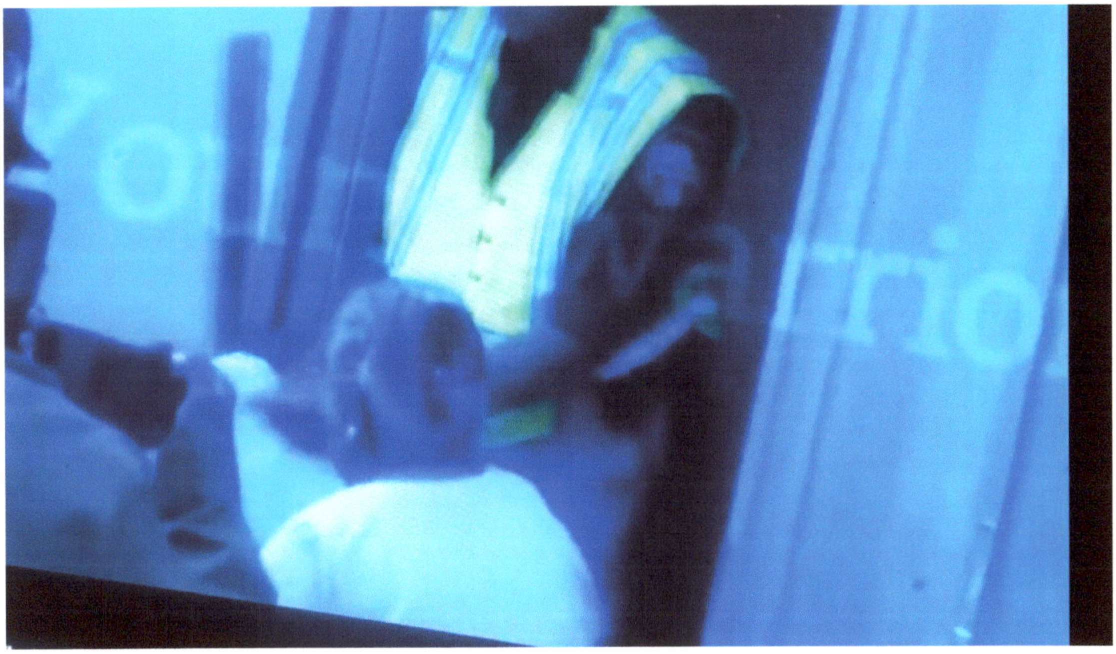

And then also for her to see how they had put the sacred fire out. Yeah, she was upset.

Robert Lundahl: So like you said, to repeat, you are leaders in the Native American Church in the State of Nevada and Native American religion is protected under the First Amendment of the Constitution, freedom of religion just like everybody else, and also their specific laws like AIRFA, American Indian Religious Freedom Act.

Why didn't this get into the heads of any of those people at the Sheriff's Office or the court that you visited and so forth. I mean, they should know right?

Dorece Sam: I don't know. Everybody- we say, they probably got paid off.

I'm not too sure, I mean I know, at the beginning when the sheriffs went out there, they didn't really want to get involved with it. I know when I was served the TRO one of the guys that handed me the TRO knew who I was, which surprised me, but he told me he didn't want to do it but he said it's his job, and I told him I said, "I understand."

Another time, I was driving down 95 and we were working on a rundown vehicle when Highway Patrol pulled up and right away, he knew who I was and he said that he supports what we're doing, trying to do our land. He said it was off the record...

Robert Lundahl: ...and meanwhile, on the record, all your sacred items are in a garbage bag somewhere, right? So what- they say you know as a community member, I live here too, or I have a family in town or those kinds of things, I support you, I know you, I'm going to see you again in the market; you know we're not- we can't never have a relationship or not talk ever again.

I mean he's there, but then at the same time all this is going on and I'm thinking about the accountability.

Does Lithium Nevada have a relationship with the County Sheriff's Department? Do they support them financially?

Dorece Sam: Well I'm sure it's happening because when they raided Camp, they had Lander County Sheriffs there, so I mean if it was just Humboldt County you know, it would have been just Humboldt County, but obviously there's some payoffs happening because it's a different County that's coming in assisting Humboldt County and the only time that ever happens is if there are drugs involved and there was no drugs involved.

So, to me, I think there is something happening there.

Chapter Eight

A Story Told: Benefits or a Curse at Thacker Pass

"We can protect the Trout and the Sage Grouse; there's more protection for those than there is for Native people." –Shelley Harjo

(LOS ANGELES, Calif.) - It wasn't much more than a month ago that I interviewed Shelley Harjo. A Paiute Shoshone mom and recently appointed Tribal Council member at Fort McDermitt, Shelley has had doubts about Lithium America's gigantic Lithium open pit mine at Thacker Pass, nearby.

Shelley Harjo

Here, we inspect the rhetoric and the reality behind America's thirst for Lithium to power batteries for electric vehicles, and rechargeables of all types from your toothbrush to mobile phones, laptops and all connected devices and tools.

Here, Shelley takes us behind the scenes to reveal the pressure her community is feeling to accept the terms of agreement handed to them by the company, requiring non-disclosure (Gag Order) in exchange for community "benefits" that may be a blessing or a curse. It is here, a story told.

Robert Lundahl: Fort McDermitt Paiute-Shoshone Indian Reservation, Nevada
The Fort McDermitt Paiute and Shoshone Tribe is a federally recognized tribe of Northern Paiute and Western Shoshone peoples, located at McDermitt, Nevada and Oregon.

The Fort McDermitt Indian Reservation spans the Nevada–Oregon border, in Humboldt County, Nevada and Malheur County, Oregon, near the Quinn River, which runs through the Tribe's Nevada lands, east to west.

Shelley Harjo: My name is Shelley Harjo. I'm an enrolled member with the Fort McDermitt Paiute Shoshone tribe. I have been involved with or watching this issue of Lithium America coming to our area.

I was recently appointed into a seat on the Tribal Council in May of this year, 2023. I had worked for the Reno Sparks Indian Colony close to 18, 19 years total.

They were involved with the protection of sacred sites because they're one of the few tribes in Nevada that has a THPO office (Tribal Historic Preservation Office), so they asked me what my thoughts were on the Thacker Pass issue. So, I gave them my thoughts, knowing it would be good to create an allyship between the tribes.

Robert Lundahl: Prescient today is the fact that by 1858 the Indians and Whites had agreed to a treaty with the principle of equal justice for all. Thieves and killers, white or Indian, were to be turned over to the authorities.

Thereafter, Paiutes under Numaga fought alongside Whites against raiding parties of Pit River Indians from across the Sierra Nevada. For the next two years, the Paiutes and Whites lived in relative peace.

However, the winter of 1858 was especially harsh, making it impossible to get provisions or people over the Sierra Nevada. Food became scarce, forcing residents to hunt for wild game. A massacre by the Pit River people had left bodies scattered across the land.

Shelley Harjo: In the meantime I had understood that the Community Benefits Agreement was started with the tribe with Lithium America and so I have always been against the mine and so I thought, okay, I need to voice my concern because politically in McDermitt, on the Tribal Council we've had a lot of ups and downs, meaning a lot of infighting.

I knew that the Thacker Pass issue was going to be a big one at some point, so when the position came open, I thought that would be my opportunity to voice the concern of some of the members who actually are against the mine.

I can give a different perspective to the sitting Council and hopefully come to an agreement on where we stand as a whole community in McDermitt. So that's kind of what's transpired previous to this.

Apparently, there's been quarterly meetings with Lithium America and so for 2023, the quarterly meeting was held August 14th, I believe, which was this past Tuesday.

Since I live in Reno it's been hard to get back and forth. Because I've planted myself here in Reno for the last 30 years, the thinking of some of the people on the council and even some some of my own people, the membership, is that once you leave, your interest doesn't necessarily follow our interests because, it's presumed that we are not in touch with what's going on on the reservation.

Robert Lundahl: In 1936 the Indian reservation was established, when the tribe organized under the Indian Reorganization Act of 1934. The reservation has 16,354 acres in Nevada and 19,000 acres in Oregon. As of the 2010 census, 313 Indians lived on the reservation, with 42 enrolled members living in nearby McDermitt.

Shelley Harjo: However, in my case, well I guess in any case, anybody who lives off of the reservation, of course, we still expect that eventually when we die we're going to be buried with our people there, so that's the the belief that us off reservation people live by.

So, I thought it was my opportunity again to get my voice out there along with the enrolled members who don't have that voice.

So, the meeting happened, and I regretfully had to inform the Council I wouldn't be there because of personal issues. This would have been my first quarterly meeting as a sitting Tribal Council member, so knowing the importance, I was very upset that I couldn't make it.

I got a few Community calls after the meeting which indicated that they're pushing forward. They want to continue, and it's the consensus of the council to continue working with Lithium America.

There were community people in the audience who were in opposition, however, you know, when we're elected officials, we are the voice of the people so with my voice not being there, that voice was left out that day.

The opposition side has always kind of just been swept under the rug because our Council believes that this is the right thing to do. You know, sign that agreement and hopefully we can get some kind of monetary assistance to help us with infrastructure on the reservation, so that's kind of the push.

We are a poor tribe. We rely on government grants and 638 contracts, so there really is no Economic Development there, as far as you know the tribe making their own money. So, it's just, in my eyes, a lose-lose situation. Because we're looking at the money, which the rest of the world is also looking at, and the benefits of "Green Energy."

We're sacrificing our future and our children's future and so that's my belief. Our people weren't heard that day.

Robert Lundahl: The Fort McDermitt Military Reservation was established 14 August 1865 at the former site of Quinn River Camp No. 33, and a stagecoach stop, Quinn River Station, in what was a traditional seasonal homeland of the Paiute, Shoshone and Bannock peoples.

Originally, the fort was established to protect the stagecoach route from Virginia City through Winnemucca, Nevada to Silver City, Idaho Territory.

It was named after Lt. Col. Charles McDermit, commander of the Military District of Nevada, who was killed in a skirmish in the area in 1865. Of several skirmishes or attacks by colonial forces including The First Nevada Cavalry in the same year, murders were between 30 and 70 predominantly women and children.

Their records and remains were recorded by the EIS for Lithium America's development/ destruction of the lands for a Lithium mine with little compensation and no direct reparations.

When the military outpost was closed in 1889, the Military Reservation was adapted as the Fort McDermitt Indian Agency, betraying its original purpose.

Shelley Harjo: In the community benefits agreement- it's a confidential agreement, so I really can't say too much other than what has been disclosed already, which was they were to build us a community building, which would assist with our childcare.

We do not have a current Child Care Program and that's one of the issues of the reservation. We have working parents. They have to leave their kids with family or sometimes they just refuse to work because they don't have childcare.

They also wanted to give us a cultural center, and then a greenhouse where we could grow our medicinal plants that they will be tearing up in the land. So, to me, you know, that's not sufficient. I've always been against the benefits agreement and its money. You can't buy our future and so that's what I see it as.

I'm speaking today on behalf of my own personal beliefs and not necessarily the Council. I was disappointed that Lithium America is pushing that they're working with the tribe to give us what we need.

It's an agreement you know but one of the disclosures in the agreement is we cannot say anything negative about Lithium America, being on the council, or it nullifies the CBA, as they call it.

I don't know if all the the people truly understand the legalities of the CBA, or the implications of having the CBA, meaning, they come in and they wave the green money in front of us, and we have no money, so they, you know, they're basically using us to to push forward their initiative of extracting minerals so I don't know if a lot of tribal members know that. So, the concern is education for the community.

I think it's more or less a gag order, just in general, do not talk about anything but just give us your green light to basically take whatever we want from your sacred land.

We can barely keep our own current buildings operating, so if they're going to build us an eight thousand square foot building that's gonna assist us, how are we gonna manage to keep a big building operating?

You know it takes a lot more than just giving us a building that's going to depreciate in five to ten years, and we're stuck with the same destruction that we were before.

I believe there was a court case in Arizona that said if the government didn't have the rights or or the company did not have the rights to the land where they supposedly are going to put the tailings, then you know it wasn't a doable project.

However, we found that the company did not have claims on that or permission from BLM, yet (Judge Miranda) Du approved it to go ahead and move it forward, so you know that's the biggest issue I see is our little tribe is having.

We don't have money for a lawyer. We don't have money to go out and find people that are going to help us in this, so I think the best option- I've heard the Chairman say, is the news media. Or, we might as well take what we can get, but I just feel like that's not the right way to go because we're talking about our grandchildren and great grandchildren's future.

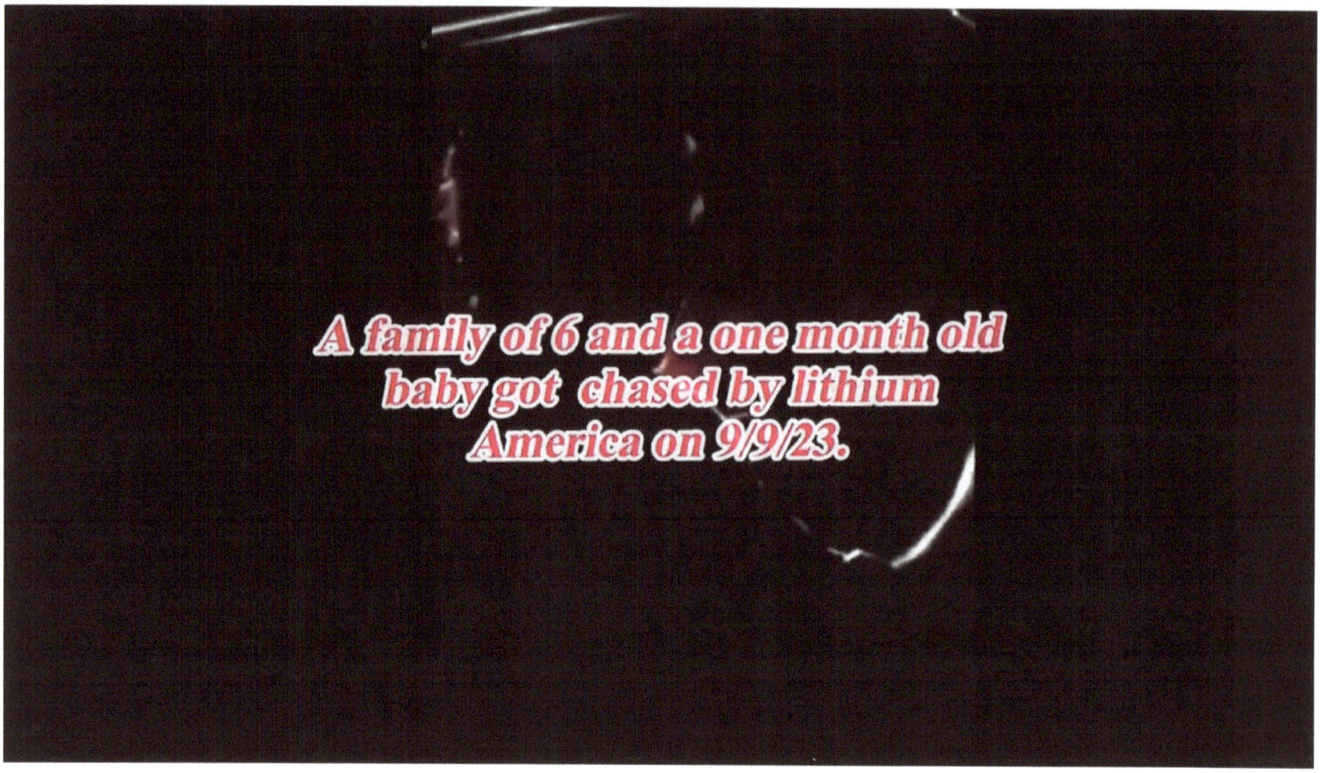

We don't know what these tailings might do, or do to us internally, as far as breathing this stuff in, all this sulfuric acid that will be in the air.

There's just so many, many reasons why we shouldn't be allowing this, and you know, for the tribes to try and fight the government over land that has always been ours, it is not right you know, we shouldn't have to.

I mean there's at some point there's a trust responsibility for the government to take care of the First Nations people and they have not. They have yet to do that.

We're the first people of this land and we have no rights when it comes to the land. I mean, it was stolen from us and the government has admitted to that yet we have no protection for even our religion, our right to religion, so I don't know how to make that point clearer.

We can protect the Trout and the Sage Grouse; there's more protection for those than there is for Native people in general. You know, it's disheartening because like I said, we're the First Nations people.

The government can protect Muslims and Jews and Christians and, you know, sexual orientation, but when it comes to Native American rights and spiritual rights and freedom to practice our religion, we're not protected.

We're what, one percent of the whole United States. They tried to get rid of us and we're still here but we're fighting the bigger battle, because you have movements like Black Lives Matter, and when it comes to MMIW (Murdered and Missing Indian Women), and Native issues, you don't hear about them, so the Inequality of Justice- it has to change.

There was a point where the government was trying to kill us off, and so I think they tried their hardest, like with the Longest Walk, there was a walk way back in, there were just so many opportunities for them to try and get rid of us, and they failed.

They even tried, and I think with this issue of Ox Sam, was when they were trying to round us up and send us on to Yakima, Washington, and Idaho and all these other places so it's always been my opinion that just because I'm on the Fort McDermitt reservation and I was born there doesn't mean that that's where my original people were.

It doesn't mean that Reno Sparks people don't belong in that area; it doesn't mean that Burns people don't belong in our area and that's what the general public doesn't understand.

We, as Paiute people, we're all over the West. So when a company like Lithium America comes in and says, we're following all our state laws and federal laws to get the approval of the people, but they only went to our tribe, and they went to our tribe when there was COVID, and I believe they sent letters in 2021, saying they didn't hear from us. Well, in July of 2021 until I would say mid-September, our tribe was going through COVID.

I had to bury my brother in August, due to COVID. He had cancer and died of COVID. He got the virus. Although he was on hospice he went home and got sick because, I believe at that time there were over 50 tribal members on the reservation who were sick.

The administration was closed. Our government was closed.

So it's always been, "I thought that the Native people, are only in their little areas of their reservation," but I think that's what the government set up for us, herd them all out like cattle and wherever they are, that's where they're going to be counted, and then they stay in their little corrals.

And I use that analogy because you know we've got a lot of ranchers in our area, a lot of the ranching mentality is in our area, and the reason why I bring this up is because when they did the walk, you know, people died all along those walks to Washington and Oregon and Idaho, and so Lithium America, this foreign company, does not understand.

Even the public doesn't understand how Native people went to the resources. We gravitated towards the resources, the land resources, based on the season.

When these companies come in, it's just, in my opinion, a modern-day genocide that's still occurring from way back when. I don't understand why we're a target, or maybe not even a target, but just- the government has always tried to get rid of the Savage Indian and I guess we're still seen as that. So, I don't, I don't have an answer for that.

It's really difficult to live in modern day society and have people understand our culture and understand why we're so connected to the land, because there's people who say, what's a rock? or what's that weed? and to us those are things that really mean a lot.

I mean we use the rocks and ceremonies, several different ceremonies, we use the medicines.

A lot of our people, when they had COVID, during that time, they went to the hills and and got the medicines to help people get over COVID, so there's a lot of things that happened and are still happening, and whether it's in our religion or traditions, our cultures; it's still happening and people don't understand that.

When you're in it, when you live it, it's just Who You Are.

When you look at Oak Flat, they're in almost the identical situation, but I think the way that theirs went, and correct me if I'm wrong, but it was a land transfer that went through on a bill.

There was a rider that had to pass and so that's how the land transfer went, but you know they've been fighting this, I guess a little bit longer than we have, but you know... what do we do?

I don't know.

Lithium America and any other foreign company mining for minerals is a foreign country and you know, I know the government says we need to stay away from other countries.

We're at the mercy of the other countries because they mine and that's where we get our minerals from. That's why they want to mine domestically but, you know, on the backs of our people.

They put us on these desolate lands thinking these are worthless areas and then when they find their mineral then you know again it doesn't mean anything to them- but it means something to us.

Whether we're there voluntarily or not, it means something to us, but to them it's, you know, they could dig seven thousand feet into the dirt and not give a world about it. They could care less.

My perspective is although I can see the benefit of an electrical vehicle, however, you know when some of these batteries overheat, there's no way to fix it. People have been locked in their cars because their batteries malfunctioned in their Teslas.

So, when you have a bad battery, how do you recycle that? I know they're working on that but how do you recycle that?

So, when you talk about electric vehicles you know you have to have an electrical grid. And if you want to push that initiative, where are we going to get all that electricity?

You're poking holes in Mother Earth whether it's for water, whether it's for lithium, whether it's for copper, but in essence what are we doing? We're poking holes in the earth that sustains our life.

I would ask people to think about the grid. Think about your grandchildren, your great-grandchildren, you know because a mineral can only last so long. Once we start digging more and more into Mother Earth, we're digging into her heart, her Water Resource that gives us life.

Life isn't going to be anymore.

We must trade something for something, you know one thing for another, life or a car.

The people of the state of Nevada who are pushing for this green energy, the people of the United States, the people of any area going through this need to think about the future. What are the pros and the benefits, because, at the end of the day, the clean energy laws, clean water, those don't really protect in the long run.

When we signed that benefits agreement we weren't thinking about the tailings in the air; we weren't thinking about maybe one of these trucks on the highway gets a flat tire and hits one of our tribal members.

I had an aunt who was killed on Highway 95 where they'll be transporting. She was killed by an RV that had a blowout. I mean we're not even thinking of those things about the protection of our people.

So those are my concerns and the reason why I'm against this mine- and there are many more to come. There are so many mining claims within the area of McDermitt it's ridiculous.

We're a small rural tribe and we have a small area that we call home, and you know, we're getting bombarded by all these foreign companies who want to come and drill in our backyard.

We've got to think outside the box, think of our grandchildren, you know, what are these tailings going to do to our lungs? They expect this to go on for 50 years and another 50 for a reclamation, and you know to me that's too long and I don't know if our lives can handle all this.

We don't know the effects until they happen. I try and be a forward thinker and think about the people.

We have a Mercury mine outside of the town of McDermitt and a lot of our people work there and a lot of them got cancer, a lot of them have died. A lot of people lost their teeth because of the sulfuric acid in the air, so I don't know if our people really understand the health risk.

We've got another mine that's come in and they're from Australia and I think they (want) an additional water well to be drilled. So that's tapping into our water. At the McDermitt reservation in the last month, and we've had a water shortage where we've had to shut down our tribal clinic because you can't see a patient if you don't have water. There are so many issues, we need to focus our efforts on something that means something, rather than money in a building.

Chapter Nine

rewilding LA: Matilija Dam History and Ecosystems

Tracker Quinones

Activist, educator and media producer, Gina "Tracker" Quinones (Chiricahua Apache), whose family once lived on tribal lands seized for the development of the Atomic Bomb, knows what it feels like to lose access to traditional lands, but also to her place-based culture, a devastating and heartbreaking loss, stripping away connections to a homeland that exists only in memory.

Tracker Quinones: I got here a couple days ago. I was in Arizona; I arrived so that I could do a report on the Matilija Dam. Yesterday there was an alert for "flash flood" warnings, so I went up there anyway, but there were road closures. I got through to the park where they had the flooding and there was a huge river in the canyon so there was no getting back there.

However, today is a little better. It stopped raining a couple hours ago but today is 93% rain forecast, and we'll probably have a over a half an inch of rain today. So we're looking at tonight, at least over a half an inch of rain again with high wind.

Robert Lundahl: Wow and so there you are in a natural system. You have rain falling into the land and into the river and going to the sea, so you're witnessing the natural process.

Tracker Quinone: Yes, and the reason why I wanted to go in yesterday is because they did draining behind the Matilija Dam and I know that there are a lot of things going on down at the bottom... things that have not been revealed or unearthed for many years.

So, it's going to be important for us to go in there and do some investigation of what's there. Now that we've had the flooding and the river pushing everything, rock, debris and life and things that are down there.

It's filling up again behind the dam so we don't know how long we'll have to wait to see, you know if the water levels are going to go down so that we can go and do some investigation.

Robert Lundahl: So you're talking about what's been described as Cultural Resources, right?

Tracker Quinones: Yes, we're talking about a history of a massacre of The Last Stand of the Chumash People, and this took place in the 1800's.

The Last Stand is important for that area back there because they had come up against that area where the hot springs are, all the way to where the lake is and the dam is and that's where a lot of the Massacre had taken place.

I have done some research looking into that particular time in history to find out if there has been any other things written about that historical landmark and I can't find very much.

We're really interested in seeing what may be there at "Ground Zero" after the water has been drained.

Now, 15, 20 years ago, they did a feasibility study by the Army Corps of Engineers and recruited us at the Mesa Matilija Environmental Science Center to be part of a logistical force and task force to be able to go back there and help to identify native species, and also to help in the study of removing the dam.

Since then, I think in 2010 they did some detonating and explosions but after that it's just always been in funding and a retrospective ideology of what should happen.

Six years ago, Candice Meneghin wrote a 170 million funding plan and a 3.3 million grant for the Matilija Dam ecosystem, and that was for a restoration program.

That was really motivating and inspiring because we are looking at a very sensitive ecosystem, and behind the dam up through the canyon there's at least four ecosystems.

I have witnessed myself the different Riparian Woodland, Wetland, and Dry Rock Bed ecosystems, and so they have recognized this as a very sensitive place that hosts a variety of animals and very sensitive place for the ecosystem to grow and thrive.

However at the same time, 20 years ago, Monsanto went back there and was spraying Glyphosate, and they were doing eradication of non-native species so we had to do a lot of propagation of the natural natives like the Willow and other plants and trees so that we can save those and the medicines back there.

There's an array of medicines and root systems, so foraging and gathering is very prevalent to the native people. We had to realize that they were spraying and over spraying a lot of the medicine, so that caused a lot of controversy.

That's why Patty Pagaling created Pesticide-Free Ojai and stopped the spraying. Since then, it has gained a lot of momentum, and she is still active, and that's who we're going to be doing the reporting with.

She still lives back there. There are many people that do live back there. Some people have gotten sick so there's a lot of controversy on that spraying and this funding the dam removal, the two bridge replacements and three levels of upgrades.

I should say new operations for the Robles diversion, and Santa Ana Bridge, so other elements are in play. She attended a program update in December and they're doing a 100%

design, and they supplement on that, so when that comes out she's doing her planning and they're doing that supportive ecosystem restoration.

Ventura County, on the other hand, is just seeing dam removal and, you know, removing of it, so that as we talked about the spawning of the salmon and things like that can happen, the life force that lives there, that thrives, can go back to its natural habitat.

At this time, it would be a disruption for them to do a fish hatchery there. It would just disrupt the whole ecosystem and the natural habitat.

Robert Lundahl: Yes, so in other words, if they were going to have a hatchery to help restore the fish stocks it wouldn't be there– I mean, you could release the fish there so that they imprint on the water and they return there but the Hatchery itself doesn't have to be at that location.

Tracker Quinone: Yes, and of course you know when the dam is removed, what happens is it connects to other fingers of tributaries, lakes, creeks and so that's going to create another vein of waterways for the **LA River**.

I think the vision would be to create a plan or a steering committee for a planning of all of the restoration of natural habitat to connect all the river systems.

Robert Lundahl: Yes, that's a vision to look at- what the river once was and to not compromise by saying, we're going to let these stocks go extinct. Can you talk about that a little bit?

What are your feelings about it and what might a tribal perspective be or the Chumash perspective?

I mean not specifically with regard to the species or science, but what does a restored ecosystem provide for the people? What does it look like?

Tracker Quinones: I am not Chumash, I am Chiricahua Apache, and I have worked in the lands of the Chumash territory, the Coastal Band and also behind Matilija in the Matilija Environmental Science Center.

I have seen an array of designs and what you can call planning for different waterways, fisheries, hatcheries but I think for a natural ecosystem at this point in time in history, of course, we can't go back and restore anything to its natural state, we can only support, introduce, and be able to encourage growth, spawning, and natural habitat.

At this time, I do know the story of the spectacular Steelhead trout that used to swim in the Los Angeles River, for instance, and of the current efforts to restore conditions for the Steelhead there.

The LA River passage and habitat structure design project is now under way, which I encourage everyone to support in any way they can, and that includes being stewards of the Arroyo Seco.

They're putting Steelhead and the LA River agenda first and foremost and they're recovering the Steelhead populations in the LA River Watershed.

Now that's sounds all pretty and nice, but you just can't put things back without also looking at the native species, plants and what has been done.

Also, when I have seen the co-existence and the bioregions of natural native plants and species cohabitate, and when the ecosystems are disturbed because of dams and pollution then we see a lot of things happen where there becomes a lot of problems, especially with pollution in the waterways.

So, as they say, "what happens upstream comes downstream" and we really need to look at the Watershed. We also need to look at what is the ecosystem at this time, because we can't go back in time and restore it to what it was naturally.

It's not going to be that way; we have to work with the native plant species that are going to actually create that photosynthesis, the natural bio rhythms of the waterways, which will support the life and habitat of the natural ecosystem of the fishery and the plants and the animals and things that come there.

I think that we have to look at, for myself and other Native peoples, we're looking at not only restoring the fishery, original fish, to be able to have a supportive habitat, but we're looking at the plants and the ecosystem, and the whole natural biorhythms of all of that together.

And I think that when we try to put dams up or when they have tried to put things in place to stop water or make water move in the direction they want they've seen that creates problems as we say downstream.

It's really important that they do consult with Native tribal land people and also there are tribal Coastal bands that have been working symbiotically with the wetlands, with the riparian lands, with the woodlands.

We know that it all is in relationship, and you can't change that. You can try to, but we know that you can't change it. It all has a relationship together.

Robert Lundahl: Thank you. That's insightful for people who have little experience or have not thought this through in any meaningful way.

So, I know you're not going to speak for the Chumash people, but in your view and experience, do they identify as "salmon people?"

Tracker Quinones: Yes, they do. They have different bands, and the Coastal Bands do. They are people who take a lot of Honor in the medicine and the waterways and the life forces that are there.

I know that they have yet to be consulted really, with the Army Corps of Engineers and the political arena of what should happen at the dam and what could happen, but I do encourage them to look at the history back there.

Look at that and take the appropriate measures so that it could be they're able to look at what the history is and how they can honor that history and then move on and be a part of the decision making regarding what's going to happen there in that area.

You have a rock quarry across the way that has been there for 30 to 40 years and that quarry is an important crucial aspect of this dam, which is just across the road.

All this quarrying activity has created its own dam and the rock, and the debris has for 40 years built up right there in that area, in the water area in the river as it flows. They have to have that vision of being able to look and see how this relationship is going to determine what's going to happen in the Waterway system.

The dam is only approximately 6 to 10 miles from the ocean, so it's not far at all. It's going to be a very highly sensitive issue when it comes to directing what they want to do, actually, and how they want to create that water system to go to the ocean because a lot of damage has been done by damming up that area.

It reminds me of the construction of the Devil's Gate Dam in 1920, and the Steelhead could no longer reach the cold waters of the upper mountain watershed of the area to spawn, and the subsequent connection of the Arroyo Seco stream and the Los Angeles River made return basically impossible.

So if you look at it today, the iron gates (the Iron Gates are all the dams, all the way up North, all the way up to Washington) have created a lot of detrimental value to not only people personally in their life to get to fish for their families, but it also it devalued the whole ecosystem along the way, including animals who would go into the water and to fish for themselves, the natural way, and the spawning, of course, of the Steelhead.

So, we know all this, but you know it still happened.

Those dams went up and now we have to live with that and and course correct it. And I'm not the one that's going to make those decisions, but I do believe that consulting the tribal people of the local tribe and First Nation people is so important.

Devil's Gate

They didn't consult them when they put the dams up, and I think now to correct the course, that would be good to have that relationship so that it can be done in the manner of relationship to all things, all life force.

Robert Lundahl: I grew up in Pasadena right by the Arroyo Seco, and there's a mandate in the State of California to teach the Native American history in third grade for all the kids.

No one told us about the Arroyo Seco and the inhabitation of the area or the the Devil's Gate Dam, and years and years later I wrote a piece called Rose Bowl Genocide.

It was published during the Rose Bowl game, and a lot of people were very interested in reading this. The area was inhabited by the Xaxaamonga people who are a band of the, who we call Tongva who are Chumash relative, essentially, very close.

I was so surprised to find out this information- that there had been a massacre there at Arroyo Seco at the Devil's Gate Dam. That it was a sacred area and a village site. This information was suppressed from my education growing up.

Xaxaamonga was the name of the Chief and the Tribe and the place. Do you have any thoughts on that? That's all from research and "book learning," you know but that's what I picked up, because I lived there, I grew up there that was my home, that was where my spirit lived when I was 5 years old, so I'm learning more about that.

Tracker Quinones: That's important, you know, those stories that you're talking about are oral stories that we value as first witness account. This is important, what you're speaking of, being able to have a relationship with the rivers. Today, that's not so, and so I do know that there has been a change– and since the dams and the Waterway systems and just even water rights.

Golden State (Water Company) came into California, and they had created contractual agreements for the water rights of California, especially Ventura County, and Ojai.

What happened was we saw a large volume of water being taken without the pipes being able to support it, so it created disaster, disaster, and emergencies after emergency. Those things are still in play and that damage is still being restored today from that volume of water being taken.

We need to look at that, like you say, of what used to be and the rivers and the waterways that we used to have access to, but no longer do.

We need to look at that perspective and say when is enough. Being able to have agricultural agreements and contracts to take that water and move it to a different county is fine but to move it to another country, I think is not okay.

It not only impacts the local waterways, but it impacts the homeowners and all of the development that has to happen that is a big problem with our system, water that is moved in large volumes.

No one, 100 years later, thought that there would be hope that we can bring some of the trout and native fish back to the Los Angeles River.

Exploring fish passage, habitat structures, and how to design stream improvements will provide good habitat and conditions for Steelhead trout and other native fish in a 4.8 Mile Stretch of LA River near downtown LA and create migration corridors in the more natural sections.

So you see, we have hope they are looking at natural corridors of water where the habitat is moving and growing and thriving naturally and to look at that before we just introduce all these trout.

Arroya Seco

We need to really see how the adaptation is going to be. It's like introducing an animal back into the wild or an eagle or a hawk back into the wild, you have to look at their predatory habitat and their practices, and their survival skills. Same with the plant, same with the fish, same with any animal.

The project objective is to enhance the native fish habitat linkages so the proposed construction will involve cutting and modifying the concrete channel that now lines the river to enhance habitat and migration corridors for native fish while maintaining flood management capacity.

So, we are 100 years later, looking at different technology or construction, or ways to create pathways for the symbiotic relationship to happen naturally, which is important after all this time.

I see people and local support having a voice.

(BLYTHE, Calif.) - Desert and Native American activists rallied Friday, June 3, at the site of the German firm Solar Millennium's Blythe Solar Project, where construction endangers several large Native American geoglyphs. The alleged destruction two large glyphs sparked the protests. Solar Millennium is a partner with Chevron Energy Solutions in the project.

Activists with the group Desert Survivors held a simultaneous rally in front of Solar Millennium's Oakland headquarters. According to spokesperson Bob Ellis, "We support the La Cuna de Aztlán Sacred Sites Protection Circle in their efforts to defend the Kokopilli intaglio and the many others on or near the project site. We favor investment and jobs for solar development in the urban areas where both the energy and the work is needed. Photovoltaics in the city are much more economical than remote, extra-expensive, obsolete, solar thermal mega-projects."

The geoglyphs in question are ancient, large drawings made on the surface of the ground. They were made by scraping away manganese-stained pebbles, to reveal lighter toned caliche below. The geoglyphs represent human-like figures and geometric forms. Their location on major trails used by multlple indigenous groups attest to their widespread cultural importance.

Although La Cuna De Aztlan Sacred Sites Protection Circle has a Memorandum of Understanding (MOU) with the Bureau of Land Management (BLM), to work together to protect geoglyphs and other cultural resources in the area, the MOU was ignored by the BLM in the permitting process, says the group's founder, Alfredo Figueroa.

Spanish Conquistador, Juan de Oñate

Indications are that the geoglyphs may be very old. According to anthropologist Jay von Werlhof et. al, in a paper titled "AMS 14C Age Constraints on Geoglyphs in the Lower Colorado River Region, Arizona and California, in Geoarchaeology: An International Journal, 1995, a representative sample of geo-glyphs along the Lower Colorado date from between AD 213 and AD 1408 with margins of error of between 57 and 392 years depending on the particular location.

The Blythe Geoglyphs are not the only cultural resources endangered by a planned "build-out" of large solar in the Mojave Desert. According to the testimony of Elizabeth A. Bagwell, Ph.D., RPA and Beverly E. Bastian before the California Energy Commission, "This analysis estimates that more than 800 sites (Cultural Resource Sites--Sacred Sites) within the I-10 corridor, and 17,000 sites within the Southern California Region will be potentially destroyed (by large solar development in application with the BLM)..."

Philip Smith, Chemehuevi

The BLM denies any geoglyphs are endangered, but La Cuna technicians indicate several remain on the project lands. Lundahl's video 'Messages in the Rocks,' specifies which sites are vulnerable. La Cuna De Aztlan's GPS coordinates for these sites are available to the press.

According to Figueroa, "The Temple" remains on proposed project lands slated for development by Solar Millenium, LLC, and permitted by the BLM.

According to La Cuna de Azlan's Chairman, Patricia Pinon, the pressure of the ARRA "fast track" process approved by Secretary of the Interior Ken Salazar resulted in inadequate Environmental Impact Statements and inadequate government to government consultation with the tribes and native groups. Bill Powers of San Diego based Powers Engineering, a solar energy consulting firm states, "So the Orwellian aspect of the government involvement in this process is that now instead of the analysis being focused on is this a cost effective project, is it a project that will minimalize or eliminate environmental damage, is it a good investment strategically for this state, all of that is secondary for the race to get the permit so that the project can get a 30 percent cash grant.

"The Temple" Geoglyph

In this case -- that's $600 million up front per project. So the money is so big that it's now like a bull rush to the finish line." In order to qualify for rebates, Solar Millenium needed to "break ground" at the site. So far, the company has graded a 100 ft. wide transmission corridor extending three miles along the path of a former gas line road.

According to La Cuna, the sacred sites destroyed include a cleared circle representing the Sun, and a second geo-glyph, known as "True North." "These images are part of our tradition of scientific inquiry predating Galileo, Copernicus, and early European astronomers," Pinon declares, "They correlate to astronomical observations."

La Cuna de Aztlan Sacred Sites Protection Circle Asks L.A. Federal Court for Halt of Solar Millennium Blythe Construction

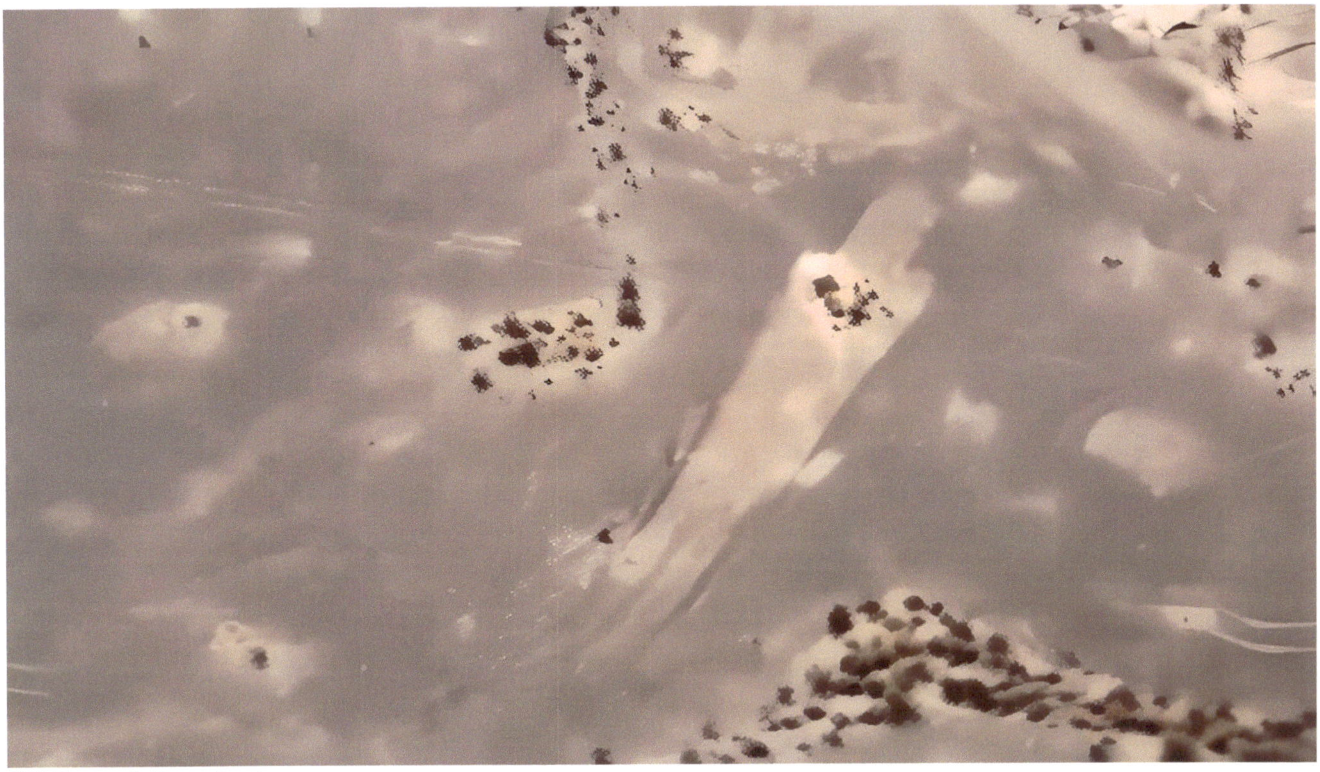

"El Tosco" Geoglyph

(BLYTHE, Calif.) - La Cuna de Aztlan Sacred Sites Protection Circle, acting to protect Native American Religious Freedom, filed a renewed application for a temporary restraining order (Case 2:11-cv-04466-JAK -OP) yesterday in United States District Court, Central District of California. According to La Cuna attorney, Cory Briggs, in the "Request for TRO" (Temporary Restraining Order), defendant Bureau of Land Management's "Right-of-Way Notice to Proceed" issued to defendant Palo Verde Solar I, LLC, on November 4, 2010, then subsequently received by the plaintiffs relating to the boundaries of the proposed Blythe Solar project, were in black and white and did not reflect actual boundary lines printed in red on the originals.

Briggs states, "It was not until early May 2011 that my firm received a copy of the Blythe Project's Notice to Proceed--nearly three months after one of my associates had to threaten the pursuit of "formal legal remedies" in order to get a copy of the Notice to Proceed. The copy that we received was black and white, and thus I was unable to discern the precise lines of the Phase I construction area authorized by the Notice. Until I received a color copy of the Notice to Proceed on June 17, 2011, I had no idea that the lines depicting the boundaries of the Phase I area were printed in red."

Briggs continues, "When I saw the red lines on the color version of the Notice to Proceed, it became clear that the access road for the Blythe Project was going through a swath of land that was not included in the environmental impact statements description of the affected land: namely, the western half of the eastern half of Section 26 of Township 6 South, Range 21 East, of the San Bernardino Meridian.

Plaintiff La Cuna has alleged that construction, which began in May at the site of the German firm Solar Millennium's Bythe Solar Project, endangers several large Native American geo-glyphs. The alleged destruction two large glyphs sparked protests recently at the site and at Solar Millennium headquarters in Oakland, CA. Solar Millennium is a partner with Chevron Energy Solutions in the project.

In the "Request for TRO," plaintiff and La Cuna Founder Alfredo Acosta Figueroa states, "I am a tribal monitor for Native American cultural sites and artifacts. There are over a dozen significant geo-glyphs or other features in or near the site of the Blythe Project, which is being challenged in this lawsuit. Several geo-glyphs and sacred features within the Project boundary will be directly impacted by the Project. Even geo-glyphs that are outside the Project's primary footprint or the footprint of the linear facilities are adversely affected because, among other things, access is impeded, air pollution and added congestion quickens the deterioration of the features, privacy necessary for certain ceremonies is lost, and construction and operation noise drones out ceremonial songs, chants, and stories that are essential to the exercise of Native American religious and cultural practices."

The Blythe Geoglyphs are not the only cultural resources endangered by a planned "build-out" of large solar in the Mojave Desert. According to the testimony of Elizabeth A. Bagwell, Ph.D., RPA and Beverly E. Bastian before the California Energy Commission, "This analysis estimates that more than 800 sites (Cultural Resource Sites--Sacred Sites) within the I-10 corridor, and 17,000 sites within the Southern California Region will be potentially destroyed (by large solar development in application with the BLM). The BLM denies any geo-glyphs are endangered, but La Cuna technicians indicate several remain on the project lands. Robert Lundahl's video "Massacre in the Rocks," specifies sites that are vulnerable.

La Cuna De Aztlan's GPS coordinates for these sites are available to all parties.

Indications are that the geo-glyphs may be very old. According to anthropologist Jay von Werlhoff et al., in a paper titled "AMS 14C Age Constraints on Geo-glyphs in the Lower Colorado River Region, Arizona and California, in Geo-archaeology: An International Journal, 1995, a representative sample of geo-glyphs along the Lower Colorado date from between AD 213 to AD 1408, with margins of error of between 57 and 392 years depending on the specifics of a particular glyph and the location.

According to La Cuna de Aztlan's Chairman, Patricia Pinon, the pressure of the ARRA "fast track" process approved by Secretary of the Interior, Ken Salazar resulted in inadequate Environmental Impact Statements and inadequate government-to-government consultation with the tribes and native groups. Bill Powers of San Diego based Powers Engineering, a solar energy consulting firm states, "So the Orwellian aspect of the government involvement in this process is that now instead of the analysis being focused on is this a cost effective project, is it a project that will minimalize or eliminate environmental damage, is it a good investment strategically for this state, all of that is secondary for the race to get the permit so that the project can get a 30 percent cash grant. In this case -- that's $600 million up front per project.

So, the money is so big that it's now like a bull rush to the finish line." In order to qualify for rebates, Solar Millennium needed to "break ground" at the site. So far, the company has graded a 100 ft. wide transmission corridor extending along the path of a former gas line road to Highway 10.

Time is weighing against the defendants. On April 6, 2011, Ronald Liebert appearing at a scheduling conference in another proceeding on behalf of Solar Millennium, LLC, told the administrative law judge that the Blythe Project still did not have all necessary approvals and that it needed them by September 2011.

"We have some other important unfortunately major time pressures. And it has to do a lot with financing the project. Of overriding concern to us is, unfortunately, by the end of September, we have to have proved up this case. We must have PUC approval. We have to go back and get Energy Commission revised permits. We must deal with BLM. We have to submit all of that, and plus we have to give DOE significant amount of time to ask questions and answer them all by September 30th. If we miss that date, there is a high likelihood that financing this project would be in jeopardy.

Whether or not that means a project could be canceled, whether or not that means the project can't get built anywhere in the time frame is really uncertain. All we know is we've got this date. And, frankly, we are scared to death right now because the government is debating whether to shut down U.S. Government and defund certain projects. And to the extent that can't get it all settled even before the 30th, we don't know what DOE will do. If funding runs out for them in mid-September, if we are not in by then, we have a big problem. So we will do anything that you require to get this in and done."

According to plaintiff and Chemeheuvi Elder Philip Smith, of Needles, CA, the loss of these sites and trails would be permanent and would irrevocably impair his ability to practice his religion. His religious practices are inextricably tied to his ability to visit and pray at the sacred sites, which provide for him the ability to make "the connection between earth, self, and spirit" that are central to his religious beliefs and practices. Regarding the pending destruction of the geo-glyphs, rock scatters and mounds, trails and other archaeological features situated within and adjacent to the boundaries of the Blythe Solar Millennium project, Smith expresses himself in the following way:

"If you erase the geo-glyphs and trails (at the Blythe Solar Millennium Site and surrounding areas,) those geo-glyphs and trails are sacred sites in my religious beliefs and practice. If you erase all of that, I lose the ability to make the connection, through prayer and communication with the sacred earth sites, while at the sacred site that is the center of my religious practice. Our church, the church of the Chemehuevi and Mojave and other Colorado River Indian tribes, is on the rocks and it is on the trails such as the one that passes in the area near the geo-glyphs. These, the sacred sites of the desert, of the earth, are where we pray and hold religious ceremonies. The Blythe Solar Millennium Project will destroy these sacred sites where we pray and practice our religion."

On December 28, 2010, La Cuna de Aztlan Sacred Sites Protection Circle filed complaints in United States District Court, Southern District of California, challenging the Bureau of Land Management permitting processes related to six large solar facilities planned for the Mohave, Sonoran and Colorado deserts of Southern California. The group was joined by CARE, Californians for Renewable Energy, and 6 individual Native American plaintiffs. Case numbers for complaints, now re-filed, are as follows: Blythe: 11-CV-04466, Ivanpah: 11-CV-00400, Chevron: 11-CV-00395, and Imperial: 10-CV-02664. Complaints referring to the Calico and Genesis projects have not yet been re-filed.

Project applicants and the Department of Energy have since been named as defendants. The suits follow the filing of the Quechan tribe's legal challenge to the imperial Valley Solar Project which was granted an injunction by Judge Larry Burns on December 15, Civil Action No. '10CV2241 LAB CAB, United States District Court for the Southern District of California.

La Cuna de Aztlan's complaint, and now its "Renewed Request for TRO" (Temporary Restraining Order), add more impediments to industrial development of the deserts of California.

According to a study commissioned by La Cuna de ;, and conducted by an MIT trained engineer, there are at least 19 ;;; and cultural resources in the immediate landscape surrounding the 9500-acre site. According to the most recent project boundaries available, at the time of the final EIS (FEIS, 08/02/10) least 5 were within project boundaries.

Regarding the sites in question," There is much to learn," Pinon adds. But generations to come may not have that opportunity unless a suit brought in federal court by La Cuna de Aztlan succeeds. On December 28, 2010, La Cuna de Aztlån Sacred Sites Protection Circle filed complaints in United States District Court, Southern District of California, challenging the Bureau of Land Management permitting processes related to six large solar facilities planned for the Mohave, Sonoran and Colorado deserts of Southern California.

"Kokopili" Geoglyph

The group was joined by CARE, Californians for Renewable Energy, and 6 individual Native American plaintiffs. Case numbers for complaints, now re-filed, are as follows: Blythe: 11-CV-04466, Ivanpah: 11-CV-00400, Chevron: 11-CV-00395, and Imperial: 10-CV-02664. Complaints referring to the Calico and Genesis projects have not yet been re-filed. Project applicants and the Department of Energy have since been named as defendants.

The suit follows the filing of the Quechan tribe's legal challenge to the imperial Valley Solar Project which was granted an injunction by Judge Larry Burns on December 15, Civil Action No. '10CV2241 LAB CAB, United State District Court for the Southern District of California. La Cuna de Aztlan's complaint adds another impediment for industrial development of the desert.

According to Figueroa, Solar Millennium's grading was an "irresponsible act" designed to "set the stage for the transfer of U.S. taxpayer rebate and loan guarantee monies." Figueroa, whose Native American lineage is both Yaqui and Chemehuevi, says he and his family are devastated by "this atrocity committed without morals, and without proper analysis."

Ometeotl, The Creator

He says that La Cuna de Aztlan Sacred Sites Protection Circle will do everything in its power to stop the ongoing "cultural genocide," which he likens to the Taliban's destruction of the 6th century Bamiyan Buddhas in Afghanistan. The Buddhas were intentionally dynamited and destroyed in March 2001 on orders from leader Mullah Mohammed Omar after being declared "Idols."

Chapter Ten

Rose Bowl Genocide

The Xaxaamonga

When I was a boy, growing up in Pasadena, California, we played and rode horses on lands and trails along the Arroyo Seco, translated as "Dry Stream," roughly the location of the Rose Bowl. The Arroyo Seco was not always dry, floods roared out of the San Gabriel Mountains, down and beyond what is now Devils Gate Dam.

The Xaxaamonga, in their native language, are a band of the Tongva people of California. Their language belongs to the Uto-Aztecan family. Xaxaamonga was the name of the Chief, the tribe and the location, where the Rose Bowl now stands.

According to Wikipedia, "above Devil's Gate, the rapids of the Arroyo Seco are positioned so that the falls make a beating, laughing sound. In Tongva traditional narratives, this is attributed to a wager made between the river and the coyote spirit."

As Penn State and USC meet to play in the 2017 Rose Bowl, it is not forgotten by this writer upon what foundation the pageantry rests.

Tongva/Chumash Pictographs, Burro Flats.

When I was a boy, growing up in Pasadena, California, we played and rode horses on lands and trails along the Arroyo Seco, translated as “Dry Stream,” roughly the location of the Rose Bowl. The Arroyo Seco was not always dry, floods roared out of the San Gabriel Mountains, down and beyond what is now Devils Gate Dam.

The Xaxaamonga, in their native language, are a band of the Tongva people of California. Their language belongs to the Uto-Aztecan family. Xaxaamonga was the name of the Chief, the tribe and the location, where the Rose Bowl now stands.

According to Wikipedia, "above Devil's Gate, the rapids of the Arroyo Seco are positioned so that the falls make a beating, laughing sound. In Tongva traditional narratives, this is attributed to a wager made between the river and the coyote spirit."

In 1769 the Sacred Expedition led by Fr. Junipero Serra began the process of establishing the mission system and initiating the Spanish colonization of Alta California. San Gabriel Archangel, the fourth mission of the twenty-one, was established in 1771.

In 1778, the mass conversions of the Indians began. As was to be the case in the rest of California under the mission system, contagious diseases of European origin took an enormous toll on the Tongva. Mission conditions were harsh, tedious, confining, unsanitary, confusing, and culturally alien. Population decline was rapid. Though this period of history is often characterized as romantic, it was an unmitigated disaster for California's Native People.

When the Spanish era ended in the early 1820s, Mexico secularized the missions. And for the Tongva/Gabrieleño, this resulted in a massacre at Las Flores Canyon near the Rose Bowl itself, just at the top of Lake Street, a posh shopping district.

According to the eyewitness account of a Californian named Philippe Lugo, Mexican forces destroyed "the greater part of them."

I was born in Pasadena on the 21st of December, a sacred day for the Tongva.

PHOTO: Arrow of Light, Teena Takata 2004

Anthropologist Al Knight has described the importance of the winter solstice to the local Chumash and related tribes as follows: "The entire local Native American Indian religious ritual cycle is centered on the moment of winter solstice. It's like rolling together our Thanksgiving, Christmas and New Year's celebration in one event.

We lived in San Marino in what had been an ex-urban area called Chapman Woods, with a little creek, near the Tongva villages of Akuuranga and Alyeupkigna. For the most part, this is hidden history. California elementary curriculum mandates the teaching of Native history, but that history for me was sanitized, romanticized and for the most part irrelevant and inaccurate, context - free. The history of the Arroyo Seco, but a mile or two from my home was not mentioned.

"Who Are My People?," (2017) led me on a journey from my youth. Beginning as a trip from the Pasadena of my childhood - the film reaches a pitched catharsis in the desert.

Here, modern-day Indigenous elders fight international corporations to preserve and protect a small part of what's left.

PHOTO: Arrow of Light, Teena Takata 2004

###

www.ingramcontent.com/pod-product-compliance
Lightning Source LLC
LaVergne TN
LVHW071242210926
848486LV00007B/325